Diet in Brain Health and Neurological Disorders

Diet in Brain Health and Neurological Disorders: Risk Factors and Treatments

Special Issue Editor

Jason Brandt

MDPI • Basel • Beijing • Wuhan • Barcelona • Belgrade

MDPI

Special Issue Editor
Jason Brandt
The Johns Hopkins University School of Medicine
USA

Editorial Office
MDPI
St. Alban-Anlage 66
4052 Basel, Switzerland

This is a reprint of articles from the Special Issue published online in the open access journal *Brain Sciences* (ISSN 2076-3425) form 2018 to 2019 (available at: https://www.mdpi.com/journal/brainsci/special_issues/diet_brainhealth_neurodisorders)

For citation purposes, cite each article independently as indicated on the article page online and as indicated below:

LastName, A.A.; LastName, B.B.; LastName, C.C. Article Title. *Journal Name* **Year**, *Article Number*, Page Range.

ISBN 978-3-03921-650-5 (Pbk)
ISBN 978-3-03921-651-2 (PDF)

Contents

About the Special Issue Editor

Jason Brandt, Ph.D. is Professor of Psychiatry & Behavioral Sciences and Professor of Neurology at the Johns Hopkins University School of Medicine. He also holds a joint appointment in the Department of Mental Health at the Johns Hopkins Bloomberg School of Public Health. Dr. Brandt's research focuses on the neuropsychology of memory and other cognitive disorders as they appear in Alzheimer's disease, Huntington's disease, and other dementia syndromes. He has also investigated the cognitive changes associated with epilepsy and its treatments, the psychological consequences of genetic testing for neuropsychiatric disorders, and improved methods for neuropsychological assessment. Dr. Brandt is a Fellow of both the American Psychological Association and the Association of Psychological Science, and is board-certified by the American Board of Clinical Neuropsychology. In 2015, he was bestowed the Distinguished Career Award by the International Neuropsychological Society.

brain sciences

MDPI

Editorial

Diet in Brain Health and Neurological Disorders: Risk Factors and Treatments

Jason Brandt

Johns Hopkins University School of Medicine, Baltimore, MD 21205, USA; jbrandt@jhmi.edu

Received: 11 September 2019; Accepted: 11 September 2019; Published: 13 September 2019

The role of nutrition in health and disease has been appreciated from time immemorial. Around 400 B.C., Hippocrates wrote: "Let food be thy medicine and medicine be thy food." In the 12th century, the great philosopher and physician Moses Maimonides wrote "any disease that can be treated by diet should be treated by no other means." Now, in the 21st century, we are bombarded by claims in the media of "superfoods," wondrous nutritional supplements, and special diets that promise to cure or prevent disease, improve health, and restore functioning. Much of the focus has been on neurological disease, brain health, and psychological functioning (behavior, cognition, and emotion).

The hyperbole aside, the past two decades have seen considerable progress in our understanding of the role of specific nutrients and dietary patterns to brain development, physiology, and functioning [1–4]. The chapters in this volume are but a sampling of the latest research on the role of specific compounds and nutrients in brain function and dysfunction, and use of diet for the prevention and treatment of neurological and psychological disorders.

The ω-3 and ω-6 polyunsaturated fatty acids (PUFAs) have long been recognized as essential to cell membranes and normal neuronal function. Deficiencies ω-3 PUFAs have been associated with everything from mood disorders to schizophrenia to Alzheimer's disease. Fuentes-Albero and colleagues [5] report that Spanish schoolchildren with attention-deficit hyperactivity disorder (ADHD) consume diets that are lower in ω-3 PUFAs than their peers without ADHD. Since this is a case-control observational study rather than a randomized trial, no causative conclusions can be drawn. However, the authors' call for increased consumption of fatty fish (the main dietary source of ω-3 PUFAs) as a component of healthy eating patterns is well supported by other research.

Another constituent of just about everyone's diet is caffeine, by far the most commonly used psychoactive substance worldwide. Ueda and Nakao [6] examined the acute effects of this stimulant on cognition and electrophysiology in a small group of healthy young men. To achieve peak blood level quickly, the drug was "vaped" rather than ingested. Caffeine produced a slightly greater increase in working memory performance (N-back task) than did placebo. EEG power in the theta band was enhanced after inhaling caffeine vapor, but only from selected right-hemisphere frontal, central, and temporal electrodes. The authors conclude that caffeine, an adenosine receptor blocker, increases the activity of right-hemisphere regions that mediate attentional and executive functions required for the working memory task. Furthermore, transpulmonary administration of caffeine resulted in a very rapid change in brain activity. Its effects on other aspects of cognition and emotional states remain to be investigated.

One of the most active areas of nutritional research is the role of dietary fats on brain functioning. Loprinzi and colleagues [7] surveyed the scientific literature on the effects of high-fat diets on learning and memory. They found that all laboratory studies of rodents found significant deleterious effects of high-fat diets compared to standard diets. However, all reviewed studies also found that having subjects engage in regular physical exercise could counteract this effect. A variety of mechanisms are proposed, including increases in BDNF and synapsin-1 and decreases in proinflammatory cytokines and insulin resistance. Whether such a beneficial effect would be seen for all varieties of human memory (e.g., episodic/semantic, declarative/procedural, retrospective/ prospective) after an exercise regimen (of what type? how vigorous? how often?) remains to be determined.

Speaking of fats, there is at least one form of high-fat diet that has clear neurotherapeutic and perhaps neuroprotective effects. As described by McDonald and Cervenka [8], the ketogenic diet, which combines large amounts of fat with very low amounts of carbohydrates, induces the liver to produce ketone bodies (acetoacetate and β-hydroxybutyrate) which are then used as a source of energy for neurons. Ketogenesis also alters the balance of excitatory and inhibitory neurotransmitters, modifies gene expression, reduces oxidative stress and inflammation, and has other effects on brain function. The ketogenic diet was introduced for the treatment of epilepsy 100 years ago, and is today a pillar in the treatment of medication-refractory seizures. It is currently being investigated for a wide variety of other neurological disorders, including stroke, glioblastoma, amyotrophic lateral sclerosis, and Alzheimer's disease.

In the final contribution to this volume, Poulimeneas and colleagues [9] tackle a vexing question in nutritional neuroscience, namely the brain mechanisms that support successful weight loss in obese people. They review the literature on functional neuroimaging in weight-loss maintainers compared to currently obese and lean individuals. Although only eight studies, using very different methods, were located, some trends were discerned. Formerly-obese people appear to display the same cerebral activation to food stimuli in reward-related brain regions as do obese people. However, they also display heightened activation in regions of the prefrontal cortex associated with inhibitory control. The authors then speculate on the biological mechanisms underlying this "neural restraint" among weight-loss maintainers. Since most studies have been cross-sectional rather than longitudinal, we do not know whether these brain activation patterns are contributors to or consequences of successful weight-loss maintenance. One way to address this might be to determine whether those who lose and maintain a great deal of weight via bariatric surgery would display the same cerebral activation patterns as those who achieved their weight loss the "old-fashioned way" (diet, exercise, and behavior modification).

This slim volume cannot do justice to the richness of the evolving neuroscience related to diet and nutrition. I suspect that the next two decades will see even more exciting advances. Stay tuned.

Conflicts of Interest: The authors declare no conflict of interest.

References

1. Gómez-Pinilla, F. Brain foods: The effects of nutrients on brain function. *Nat. Rev. Neurosci.* **2008**, *9*, 568–578. [CrossRef] [PubMed]
2. Zamroziewicz, M.K.; Barbey, A.K. Nutritional cognitive neuroscience: Innovations for healthy brain aging. *Front. Mol. Neurosci.* **2016**, *10*, 1829. [CrossRef] [PubMed]
3. Oleson, S.; Gonzales, M.M.; Tarumi, T.; Davis, J.N.; Cassill, C.K.; Tanaka, H.; Haley, A.P. Nutrient intake and cerebral metabolism in healthy middle-aged adults: Implications for cognitive aging. *Nutri. Neurosci.* **2017**, *20*, 489–496. [CrossRef] [PubMed]
4. Sandhu, K.V.; Sherwin, E.; Schellekens, H.; Stanton, C.; Dinan, T.G.; Cryan, J.F. Feeding the microbiota-gut-brain axis: Diet, microbiome, and neuropsychiatry. *Transl. Res.* **2017**, *179*, 223–244. [CrossRef] [PubMed]
5. Fuentes-Albero, M.; Martínez-Martínez, M.I.; Cauli, O. Omega-3 long-chain polyunsaturated fatty acids intake in children with attention deficit and hyperactivity disorder. *Brain Sci.* **2019**, *9*, 120. [CrossRef] [PubMed]
6. Ueda, K.; Nakao, M. Effects of transpulmonary administration of caffeine on brain activity in healthy men. *Brain Sci.* **2019**, *9*, 222. [CrossRef] [PubMed]
7. Loprinzi, P.D.; Ponce, P.; Zou, L.; Li, H. The counteracting effects of exercise on high-fat diet-induced memory impairment: A systematic survey. *Brain Sci.* **2019**, *9*, 145. [CrossRef] [PubMed]

8.   McDonald, T.J.W.; Cervenka, M.C. The expanding role of ketogenic diets in adult neurological disorders. *Brain Sci.* **2018**, *8*, 148. [CrossRef] [PubMed]

9.   Poulimeneas, D.; Yannakoulia, M.; Anastasiou, C.A.; Scarmeas, N. Weight loss maintenance: Have we missed the brain? *Brain Sci.* **2018**, *8*, 174. [CrossRef] [PubMed]

Article

Omega-3 Long-Chain Polyunsaturated Fatty Acids Intake in Children with Attention Deficit and Hyperactivity Disorder

Milagros Fuentes-Albero [1], María Isabel Martínez-Martínez [2] and Omar Cauli [2,*

[1]   Children's Mental Health Center, Hospital Arnau de Villanova, 46015 Valencia, Spain;
      milagrosfuentesalbero@yahoo.es
[2]   Department of Medicine and Nursing, University of Valencia, 46010 Valencia, Spain;
      m.isabel.martinez@uv.es
*   Correspondence: omar.cauli@uv.es; Tel.: +34-96-386-41-82

Received: 15 April 2019; Accepted: 22 May 2019; Published: 23 May 2019

Abstract: Omega-3 long-chain polyunsaturated fatty acids (LC-PUFA) play a central role in neuronal growth and in the development of the human brain, and a deficiency of these substances has been reported in children with attention deficit hyperactive disorder (ADHD). In this regard, supplementation with omega-3 polyunsaturated fatty acids is used as adjuvant therapy in ADHD. Seafood, particularly fish, and some types of nuts are the main dietary sources of such fatty acids in the Spanish diet. In order to assess the effect of the intake of common foods containing high amounts of omega-3 polyunsaturated fatty acids, a food frequency questionnaire was administered to parents of children with ADHD ($N = 48$) and to parents of normally developing children (control group) ($N = 87$), and the intake of dietary omega-3 LC-PUFA, such as eicosapentaenoic acid (EPA) and docosahexaenoic acid (DHA), was estimated. Children with ADHD consumed fatty fish, lean fish, mollusks, crustaceans, and chicken eggs significantly less often ($p < 0.05$) than children in the control group. The estimated daily omega-3 LC-PUFA intake (EPA + DHA) was significantly below that recommended by the public health agencies in both groups, and was significantly lower in children with ADHD ($p < 0.05$, Cohen's d = 0.45) compared to normally developing children. Dietary intervention to increase the consumption of fish and seafood is strongly advised and it is especially warranted in children with ADHD, since it could contribute to improve the symptoms of ADHD.

Keywords: fish intake; omega-3 fatty acids; nutrients; ADHD; children; diet-deficient

---

1. Introduction

There is a growing evidence that several mental disorders, although they show an underlying genetic predisposition [1], are probably the product of an interplay between genetic susceptibility and environmental factors [2], of which inadequate nutrition may be a component [3,4]. Among the nutrients that have been consistently shown to be related to mental health and to different psychiatric disorders, mention must be made of omega-3 long-chain polyunsaturated fatty acids (LC-PUFA) [5–7]. A proper physical and mental health and neurodevelopment require a balanced ratio of omega-3 to omega-6 polyunsaturated fatty acids, but the typical diet in many countries provides a much larger intake of food containing omega-6 as compared to omega-3 LC-PUFA, thus often resulting in an imbalance and deficient omega-3 intake [5,8]. The consumption of supplements containing omega-3 LC-PUFA has been shown to be an effective measure in addition to the administration of psychotropic drugs for treating several psychiatric diseases [9–12]. In this regard, it has been demonstrated that omega-3 LC-PUFA such as eicosapentaenoic acid (EPA) and docosahexaenoic acid (DHA) may be helpful in the treatment of attention deficit hyperactive disorder (ADHD) in children [13–17]. Whether the pathophysiology of ADHD may be linked to inadequate bioavailability of omega-3 LC-PUFA, and whether it may be counteracted by dietary supplementation or increased intake of foods containing large amounts of omega-3 LC-PUFA, has gained growing interest in part due to the increasing knowledge of the role of nutrition in psychiatric disorders and in ADHD [18–20]. Dietary guidelines recommend regular fish consumption in all age ranges as the main source of omega-3 LC-PUFA intake [21].

Previous studies refer to the fundamental role afforded by omega-3 LC-PUFA in several essential metabolic functions, given their implication in diverse neuronal processes, as well as in cell growth, the function of cell membrane, hormonal, and immunological cross-talk, and gene expression regulation [8,22,23]. Alteration of some of these functions has been implicated in the physiopathology of ADHD [24]. Several experimental studies suggest that deficiencies of omega-3 LC-PUFA strongly alter brain function, not only during the developmental stages, but also throughout life [25]. There is some evidence to suggest that omega-3 LC-PUFA homeostasis may be impaired in patients with ADHD as a result of deficits and/or imbalances in nutritional intake, genetic alteration, changes in the activity of the enzymes involved in their metabolism, or the influence of some environmental agents [24,25].

Although many studies on omega-3 LC-PUFA supplementation in ADHD have been published in recent years [13,14], most refer to either interventions performed in patients who were given omega-3 LC-PUFA supplements apart from their normal diets. Remarkably, there are few studies on the intake of omega-3 LC-PUFA through diet in patients with ADHD. The present study was therefore designed with the following three main objectives:

1)   Evaluation of the pattern of consumption of the main dietary sources of food containing omega-3 LC-PUFA in children with ADHD and in a control group.
2)   Estimation of the daily intake of omega-3 LC-PUFA (EPA + DHA) in the two groups.
3)   Evaluation of the influence of age, sex, and body mass index (BMI) upon omega-3 LC-PUFA intake.

2. Materials and Methods

2.1. Study Design

An observational case-control study was carried out in Valencia (Spain) in 2016–2017. The study participants were recruited among patients (children and adolescents) with ADHD undergoing child psychiatrist consultation. Neurologically healthy children (control group) were recruited from two public schools in Valencia (Spain). Attention deficit hyperactive disorder was confirmed based on the DSM-IV diagnostic criteria using a standard neurodevelopment examination and interview (Conners scale). The parents of children with ADHD were interviewed during ordinary consultation with the child psychiatrist. Clinical information (diagnosis of ADHD, medication, presence of other comorbidities, anthropometric data) was retrieved by reviewing the medical records in the psychiatrist consultation of children with ADHD. Body mass index was calculated as weight in kilograms divided by the square of height in meters. For children and adolescents, BMI is age- and sex-specific, and is often referred to as BMI-for-age. According to the international guidelines, BMI was grouped into four categories: underweight (BMI less than the 5th percentile), normal or healthy weight (5th percentile to less than the 85th percentile), overweight (85th percentile to less than the 95th percentile), or obese (equal to or greater than the 95th percentile) [26].The children in the control group were sex- and age-matched (proportion 1:2) with the children in the ADHD group. Matching increases the efficiency of the estimates if the matching variables are associated with both the disease and exposure. The study comprised 135 children: 48 with a diagnosis of ADHD (age 5–14 years) and 87 with no ADHD or other psychiatric or neurological disorders (age 4–13 years). Socio-economic variables were measured through three variables: First, occupational social class, widely used in Spain as a measure of socioeconomic position [27]; it was defined using a Spanish adaptation of the British social class classification. In this study, we recoded the social status in three categories: higher, medium and lower. Educational level was recorded as primary or less, secondary, or university. Employment situation was categorized as employed, unemployed, and homemaker.

The study protocol was approved by the local Ethics Committee of the University of Valencia (Valencia, Spain) (protocol number H1397475950160). Parents signed the informed consent in order to participate in the study.

2.2. Diet Assessment

The parents completed the food frequency questionnaire (FFQ) about their children's diet and were also instructed to report all beverage and supplement consumption. The instrument was a semi-quantitative food questionnaire that was comprised of 136 food items, and is validated in Spain [28]. Specifically, the parents were instructed to record estimated portion sizes for each item ingested according to a previously validated [29] visual guide to improve the accuracy of their estimates. Consumptions were assessed by crossing the frequency and the portion size for each food. All food records were analyzed using Nutrition Data Systems-Research free software (DIAL®). Nutrient intake was averaged across the three days and normalized to intake per 1000 kcal, to generate the measures used in subsequent analyses. Energy and nutrient intake was calculated from the Spanish food composition tables [30,31].

2.3. Estimation of Omega-3 LC-PUFA Intake from Fish and Nuts

Parents self-reported fish and nuts consumption in their children. Fish was defined as "any kind of fish, including fish sticks and canned tuna fish, shellfish, crustaceans and mollusks." Participants reported: (a) how often they consumed fish ("did not eat," "once–three times a month," "about once a week," "twice–four times a week," "five–six times a week," "once a day," "twice–three time a day"); and (b) the type of fish they typically consumed.

The items of the three-day semi-quantitative food questionnaire [30] related to fish and seafood consumption and their omega-3 LC-PUFA contents (g/100 g of food item, as the sum of EPA + DHA) were: (a) lean fish: young hake, hake, sea bream, grouper, and sole (0.62); (b) fatty fish: salmon, mackerel, tuna, Atlantic bonito, and sardine (1.87); (c) cod (0.70); (d) smoked and salted fish: salmon and herring (4.44); (e) shellfish: mussel, oyster, and clam (2.20); (f) seafood: shrimp, prawn, and crayfish (0.90), and (g) mollusks: octopus, cuttlefish, and squid (0.71). Omega-3 LC-PUFA intake was calculated as frequency × (EPA + DHA) content for each food item (fish, seafood). We also included common foods in Spanish diets containing high amounts of omega-3 LC-PUFA such as dry fruit nuts: walnuts, hazelnuts, and almonds (6.33) [28,32]. We estimated the intake of EPA + DHA because these fatty acids are administered as nutritional supplements in clinical settings for children/adolescents with ADHD. In addition, we asked the parents about the frequency of consumption of omega-3 LC-PUFA supplements or omega-3 fatty acid-enriched milks. The intake of omega-3 LC-PUFA and fish consumption were adjusted for total energy intake using the residuals method proposed by Willett et al. [33].

2.4. Statistical Analysis

In the univariate analysis, variables were represented as absolute frequencies and percentages for categorical variables, and as the mean ± standard deviation (SD) for continuous (quantitative) variables. In the bivariate analysis, we first checked for normal or non-normal data distribution for quantitative variables using the Shapiro–Wilk ($n < 50$) or Kolmogorov–Smirnoff ($n \geq 50$) tests. As a result of non-normal data distribution, we used nonparametric tests, e.g., the Mann–Whitney U-test (when comparing quantitative variables between two groups) or the Kruskal–Wallis test (when comparing quantitative variables among three or more groups). Correlation analysis between quantitative variables was performed with the nonparametric Spearman test. In order to control the effect of intervening variables, partial correlations were performed. Differences between categorical variables were evaluated with the chi-squared test. In the case of food frequencies, we applied the z-test for differences between proportions to determine which of the five to seven categories differed between the control and ADHD groups. To quantify the effect size for two groups comparison we calculated Cohen's d. Statistical significance was considered to be $p < 0.05$. The SPSS version 24.0 statistical package (SPSS, Inc., Chicago, IL, USA) was used throughout.

3. Results

3.1. Description of the Sample

The characteristics of the study sample are shown in Table 1. Since ADHD shows a clear male predominance over females of about 3:1 to 4:1 in community-based samples of young individuals [1,2,34], we attempted to mimic the difference in sex distribution in our study: females in the ADHD group represented 25.0%, versus 28.7% in the control group. There were no significant differences between the groups regarding sex distribution ($p = 0.64$) or mean age ($p = 0.86$). Regarding the weight distribution of the subjects, 18.5% ($n = 25$) of the sample had low weight (percentile < 5), 36.3% ($n = 49$) showed normal weight (percentile 5–84), 23.7% ($n = 32$) were overweight (percentile 85–94), and 21.5% ($n = 29$) were obese (≥95 percentile). Significant differences in weight distribution were observed between the control and ADHD groups ($p < 0.0001$). In relation to BMI, low weight was significantly more prevalent in the control group compared to the ADHD group ($p < 0.0001$), while obesity was significantly more frequent in the ADHD group compared to the control group ($p < 0.001$) (Table 1).

Table 1. Characteristics of the study sample.

Variable	Control	ADHD	*p*-Value
Age	10.00 ± 0.27 (range 4–13)	9.54 ± 0.31 (5–14)	$p = 0.86$ (Mann–Whitney test)
Sex	Female $n = 25$ Male $n = 62$	Female $n = 12$ Male $n = 36$	$p = 0.64$ (Chi-squared test)
BMI	18.69 ± 0.39 (range 10.65–30.44)	20.89 ± 0.44 (range 15.50–28.31)	$p = 0.04$ (Mann–Whitney test)
Low weight	26.4%	4.2%	$p < 0.001$ (Chi-squared test)
Normal weight	36.8%	35.4%	
Over weight	25.3%	20.8%	
Obesity	11.5%	39.6%	
Social class	Higher: 26.4% Medium: 55.2% Lower: 18.4	Higher: 31.3% Medium: 52.1% Lower: 16.6%	$p = 0.88$ (Chi-squared test)
Employment situation	Father Employed: 97.7% Unemployed: 2.3% Mother Employed: 50.6% Unemployed: 16.1% Homemaker: 33.3%	Father Employed: 95.8% Unemployed: 4.2% Mother Employed: 58.3% Unemployed: 10.4% Homemaker: 31.3%	$p = 0.95$ (Chi-squared test) $p = 0.78$ (Chi-squared test)
Educational level	Father Primary school: 23.0% Secondary school: 54% University: 23.0% Mother Primary school: 17.2% Secondary school: 56.4% University: 26.4%	Father Primary school: 20.8% Secondary school: 54.2 University: 25.0% Mother Primary school: 12.5% Secondary school: 56.2% University: 31.3%	$p = 0.89$ (Chi-squared test) $p = 0.84$ (Chi-squared test)

No significant differences in the socio-economic variables were observed between parents in the ADHD and control group such as social class, employment situation, and educational level (Table 1).

3.2. Energy Intake and Frequency of Seafood Consumption

The reported average energy intake was approximately 1705 kcal. Of this amount, 51% corresponded to carbohydrates, 34% to fat, and 15% to protein. Fish intake was significantly lower in children/adolescents with ADHD than among the controls for all types of fish and seafood, except codfish. Significant differences were recorded in relation to lean fish (including young hake, hake, blackspot sea bream, goliath grouper, and common sole) ($p < 0.001$) (Figure 1A); the z-scores analysis showed significant differences for the intake categories "once a week" (z-score = 3.05; $p < 0.01$, higher in the control group), "twice–four times a week" (z-score = 2.15; $p < 0.05$, higher in the control group) and "five–six times a week" (z-score = −4.14; $p < 0.001$, higher in the ADHD group). Significant differences were also observed in the case of fatty fish (salmon, mackerel, tuna, bonito, sardine) ($p < 0.001$; chi-squared test) (Figure 1B); the z-scores analysis showed significant differences for the intake categories "did not eat" (z-score = −3.77; $p < 0.001$, higher in the ADHD group), "once–three times a month" (z-score = −3.54; $p < 0.001$, higher in the ADHD group), "once a week" (z-score = 2.55; $p < 0.01$, higher in the control group), and "twice–four times a week" (z-score = 4.39; $p < 0.001$, higher in the control group).

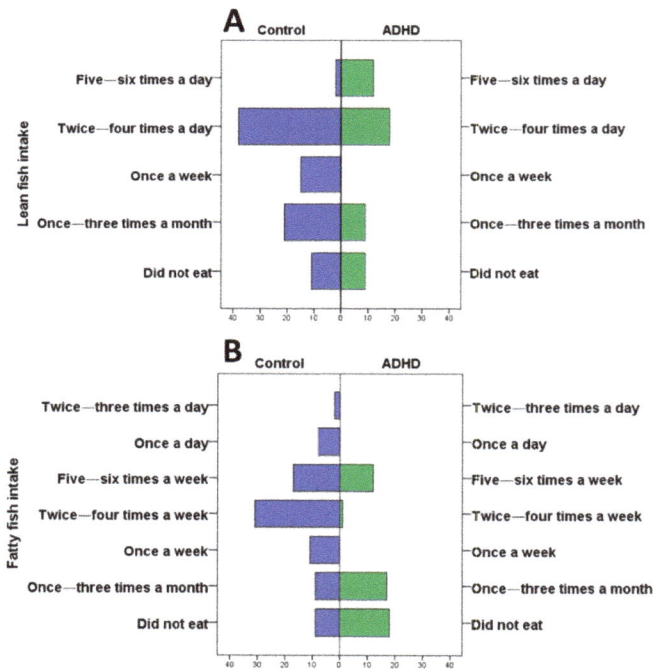

Figure 1. Frequency of intake of lean and fatty fish (Supplementary Tables S1 and S2 for raw).

Significant differences were recorded in the intake of smoked fish (including smoked and salted fish such as salmon and herring) ($p < 0.001$; chi-squared test); the z-scores analysis showed significant differences for the intake category "once a week" (z-score = 2.17; $p < 0.05$, higher in the control group). The same applied to the intake of shellfish (including mussel, oyster, and clam) ($p < 0.05$); the z-scores analysis showed significant differences for the intake category "twice–four times a week" (z-score = 2.02; $p < 0.05$, higher in the control group). Likewise, significant differences were observed in the intake of mollusks (including octopus, common cuttlefish, and squid) ($p < 0.001$); the z-scores analysis showed significant differences for the intake category "once a week" (z-score = 2.82; $p < 0.01$, higher in the control group). Lastly, significant differences were recorded in the intake of crustaceans (including shrimps, prawn, and crayfish) ($p < 0.01$; chi-squared test); the z-scores analysis showed significant differences for the intake categories "once a week" (z-score = 2.02; $p < 0.05$, higher in the control group) and "five–six times a week" (z-score = −2.11; $p < 0.05$, higher in the ADHD group). In contrast, the intake of codfish was not significantly different between the two groups ($p = 0.23$).

There were no significant differences in relation to the consumption of nuts (referred to those containing higher amounts of omega-3 LC-PUFA, such as walnuts and almonds) ($p = 0.07$), omega-3 LC-PUFA supplements ($p = 0.26$), or omega-3 fatty acid-enriched milk ($p = 0.14$). The intake of omega-3 fatty acids from nuts were not included in the calculation of daily EPA + DHA intake, since these foods contain other omega-3 LC-PUFA different from DHA and EPA, and because no significant differences in the intake of dry fruits were observed between the ADHD and control groups.

Significant differences in food intake were observed between females and males in relation to fatty fish and shellfish (being higher in males compared to females, $p < 0.05$), and eggs (again being higher in males compared to females, $p < 0.01$), but not to other foods ($p > 0.05$).

3.3. Estimation of Omega-3 LC-PUFA (EPA + DHA) Intake

The estimated ingestion of omega-3 LC-PUFA (EPA + DHA) in the diet was 109.87 ± 80.27 mg/day for the control group and 78.42 ± 56.64 mg/day for the children with ADHD ($p < 0.01$, effect size Cohen's d = 0.45) (Figure 2). The analysis of the mean intake per day of omega-3 LC-PUFA for each type of fish and seafood is shown in Table 2. There is a significant effects in omega-3 LC-PUFA between the two groups for lean fish ($p < 0.05$), fatty fish ($p < 0.01$), mollusks ($p < 0.05$), and other types of fish and seafood less frequently consumed ($p < 0.05$).

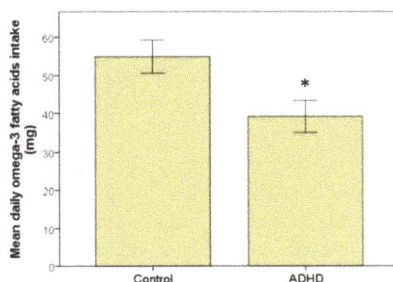

Figure 2. Estimated daily EPA + DHA intake from seafood. Comparison of EPA + DHA intake in the control and ADHD groups. Significant difference reported with an asterisk *, $p < 0.05$.

Table 2. Estimation of Omega-3 LC-PUFA intake form different type of fish and seafood.

Group	Lean Fish (mg/day)	Fatty Fish (mg/day)	Mollusks (mg/day)	Crustaceans (mg/day)	Other Types (mg/day)
Control	45.56 ± 19.81	40,63 ± 33.6	18.28 ± 18.20	3.21 ± 6.22	2.20 ± 5.11
ADHD	38.51 ± 19.22 *	26.42 ± 20.30 **	10.21 ± 15.4 *	3.0 ± 6.43	0.29 ± 2.62 *

*, $p < 0.05$; **, $p < 0.01$ compared to the control group. On considering the daily intake related to weight category, a significant difference was seen to persist between the control and ADHD groups ($p < 0.01$).

There was a significant correlation between the mean daily intake of EPA + DHA and the frequency of intake of fatty fish (rho = 0.18, $p < 0.05$) and crustaceans (rho = 0.17, $p < 0.05$). No significant differences were observed in the estimated daily amounts of omega-3 LC-PUFA (EPA + DHA) between sexes ($p = 0.17$) or among children in the different weight categories ($p = 0.57$).

In contrast, a significant and direct correlation was observed between the intake of omega-3 LC-PUFA and of the age of the children (rho = 0.21, $p < 0.05$; Spearman test). The correlation between mean daily omega-3 LC-PUFA intake and age no longer proved significant ($p > 0.05$, partial correlation) after controlling for the intervening variables, e.g., group, sex, and weight categories, suggesting that these contribute significantly to the association between omega-3 LC-PUFA intake and age.

4. Discussion

Nowadays, several studies showing that food is not only useful for providing energy for bodily functions [35], but it can also prevent or moderate several diseases and a proper diet can improve both physical and mental health [4,5,25,36–38]. Omega-3 LC-PUFA supplementation has been shown to produce beneficial effects in children with ADHD, as summarized by two recent meta-analyses, although some conflicting results have been also reported [12–15]. To our knowledge, no studies have explored whether the intake of the omega-3 LC-PUFA EPA and DHA (expressed as mg/day) through the diet is adequate in children with ADHD. The European Food Safety Authority (EFSA) recommends an average EPA + DHA intake of 250 mg/day in the pediatric population [21]. The Food and Agriculture Organization (FAO)/World Health Organization (WHO) [39] recommends an intake of EPA + DHA about 100–200 mg/day for children aged 2–6 years and 200–250 mg/day from age 6 years onwards. Our study shows worrying results in the form of a low intake of EPA + DHA in both the control group and the ADHD group compared to the amount recommended by the public health organizations (50%–60% reduction with respect to the recommended daily dose) [21,39–41]. Similar findings have also emerged from a recent French population-based study in children (3–10 years) and adolescents (11–17 years) [42]. The mean daily intake of EPA + DHA correlated significantly to age, though on correcting for BMI, which also increases with age, we still observed a low intake of these essential molecules. Interestingly, a lower intake of omega-3 LC-PUFA has also been recently reported in children with autism spectrum disorder [43] suggesting it may be a general nutritional problem affecting the pediatric population rather than a problem conditioned by some specific neuropsychiatric disorder. However, it must be pointed out that the consequences of an omega-3 LC-PUFA (EPA + DHA) deficient diet may have even worse deleterious effects in children with neurodevelopmental disorders, taking into account that omega-3 LC-PUFA supplementation has been shown to afford beneficial effects when added to the pharmacological treatment of ADHD [13–15,38,40]. The analysis and the evolution of ADHD symptoms in children with low versus normal omega-3 LC-PUFA intake deserves future investigations in order to assess its role in ADHD symptomatology. Besides omega-3 LC-PUFA, reduced fish intake could lead to other nutrients deficiencies, such as phospholipids, the neuromodulator amino acid taurine, high-quality source of protein, and beneficial marine carotenoids such as astaxanthin [44], which have been demonstrated to possess anti-oxidant properties and anti-inflammatory effects [44–48], and regular fish intake reduces hyperlipidemia [49], which in turn can improve brain function [50].

Another emerging nutritional concern in our study was the high prevalence of obesity in children with ADHD (40% of the sample compared to 12% of the control group). This finding agrees with those of a recent meta-analysis concluding that the prevalence of obesity in ADHD is 40% higher than in the general population [51,52]. The causes of being overweight and child obesity are multifactorial (diet, sedentary lifestyle, socioeconomic status, disease conditions, neurodevelopmental disorders, etc.), but the core symptoms of children with ADHD might contribute to such increased rates [43,51,52]. Among these factors, ADHD symptoms, such as inattention or impulsivity, can increase the risk of obesity by increasing and dysregulating the food intake pattern in several ways (excessive eating, binge eating, unhealthy food choices, etc.) [51]. Attention deficit disorder may be associated with not remembering whether eating has been done or with a lack of satiation feeling [51,52]. Given the lack of planning and self-regulation skills, the patient can lose control over food and reduce the time spent doing physical exercise [40]. Also, impulsivity may contribute to excessive food intake in ADHD, even in the form of binge eating. This anomalous eating pattern could produce a net increase in adipose tissue, which affects the severity of ADHD and vice versa. Bowling et al. [43] concluded that more ADHD symptoms predict higher fat mass at later ages, which further confirms that more symptoms of impulsivity may contribute to being overweight. Longitudinal studies have explored the direction

Brain Sci. **2019**, *9*, 120

of the link between ADHD and obesity. Some studies suggested that ADHD precedes, and likely contributes to, subsequent overweightness and obesity [51,52]; however, the reverse pattern has also been demonstrated in preschool children [53]. One of the proposed pathophysiological mechanism by which being overweight may contribute to ADHD relates to sleep-disordered breathing [54], leading to an excessive daytime sleepiness, which in turn may promote inattention via hypoxemia, which in turn contributes to altered prefrontal functioning [52,54]. Finally, a common genetic mechanism between ADHD and obesity has been also proposed [55]. Although the mechanism underlying the association is still unknown, preliminary evidence suggests the role of the dopaminergic reward system [56] or melanocortin system [57]. It is indeed possible that bidirectional pathways are likely involved.

Our study has a number of limitations. First, the cross-sectional observational design involved limits regarding inferences about causality between insufficient intake of omega-3 LC-PUFA and the worsening of ADHD symptoms. Second, there were a number of issues related to the completion of records. Data referred to intake may contain errors due to inaccuracies in recorded quantities and they are based on parents' reports rather than children's. However, we are confident that the self-reported information provided by parents about the nutritional assessment of their children was adequate because they showed interest in the study and they received training and support in filling out the food records. Furthermore, the attrition rate was low. We, therefore, think that the study has a good internal validity. A third limitation is the fact that we did not measure the intake of omega-3 LC-PUFA coming from other sources. Nevertheless, we are confident about the main role of fish and seafood as the principal source of EPA + DHA in the Spanish diet [36,39].

In our sample of children with ADHD and controls (age- and sex-matched with the ADHD subjects), there were considerably more boys than girls (reflecting the characteristic sex ratio observed in ADHD [1,2]), which can rule out a proper analysis for the effects of sex. Both the controls and the ADHD children were recruited not only from the same age group but also from the same geographical region, and had a similar socioeconomic status. Data were collected over the same time period (winter), and this homogeneity reduced potential sources of bias.

Despite these limitations, our study underscores the need for greater attention to the education of parents and children regarding healthy dietary habits in Spain, and as such, education is the most promising and practical complementary management strategy in ADHD. Given that fish consumption is the main source of dietary omega-3 LC-PUFA [58], interventions promoting fish consumption in a balanced diet, as well as other positive eating behaviors, are strongly warranted in the future.

5. Conclusions

The intake of seafood in particular fish, is reduced in children with ADHD compared to typically developing children and this may contribute to reduced intake of some omega-3 LC-PUFA such as EPA and DHA, essential nutrients for a proper brain development and function. Further research is required to clarify associations between ADHD symptomatology, eating patterns and health status.

Supplementary Materials: The following are available online at http://www.mdpi.com/2076-3425/9/5/120/s1, Table S1: Frequency of lean fish intake, Table S2: Frequencies of fatty fish intake.

Author Contributions: Conceptualization, M.F.-A., M.I.M.-M., O.C.; Methodology, M.F.-A., M.I.M.-M.; Formal Analysis, M.F.-A., O.C.; Data Curation M.F.-A., M.I.M.-M.; Writing—Original Draft Preparation, M.F.-A., O.C.; Writing—Review and Editing, M.F.-A., M.I.M.-M., O.C.

Funding: This research received no external funding.

Acknowledgments: We express our sincere thanks to all the parents for their time, interest, and goodwill, and all the staff involved in the studies.

Conflicts of Interest: The authors declare no conflict of interest.

Brain Sci. **2019**, *9*, 120

References

1. Sciberras, E.; Mulraney, M.; Silva, D.; Coghill, D. Prenatal Risk Factors and the Etiology of ADHD—Review of Existing Evidence. *Curr. Psychiatry Rep.* **2017**, *19*. [CrossRef]
2. Nigg, J.; Nikolas, M.; Burt, S.A. Measured Gene by Environment Interaction in Relation to Attention-Deficit/Hyperactivity Disorder (ADHD). *J. Am. Acad. Child Adolesc. Psychiatry* **2010**, *49*, 863–873. [CrossRef]
3. Sarris, J.; Logan, A.C.; Akbaraly, T.N.; Amminger, G.P.; Balanzá-Martínez, V.; Freeman, M.P.; Hibbeln, J.; Matsuoka, Y.; Mischoulon, D.; Mizoue, T.; et al. International Society for Nutritional Psychiatry Research. Nutritional medicine as mainstream in psychiatry. *Lancet Psychiatry* **2015**, *2*, 271–274. [CrossRef]
4. Yan, X.; Zhao, X.; Li, J.; He, L.; Xu, M. Effects of early-life malnutrition on neurodevelopment and neuropsychiatric disorders and the potential mechanisms. *Prog. Neuro Psychopharmacol. Boil. Psychiatry* **2018**, *83*, 64–75. [CrossRef]
5. Gow, R.V.; Hibbeln, J.R. Omega-3 Fatty Acid and Nutrient Deficits in Adverse Neurodevelopment and Childhood Behaviors. *Child Adolesc. Psychiatr. Clin. Psychiatry* **2014**, *23*, 555–590. [CrossRef]
6. Grosso, G.; Galvano, F.; Marventano, S.; Malaguarnera, M.; Bucolo, C.; Drago, F.; Caraci, F. Omega-3 Fatty Acids and Depression: Scientific Evidence and Biological Mechanisms. *Oxidative Med. Cell. Longev.* **2014**, *2014*, 1–16. [CrossRef]
7. Parletta, N.; Milte, C.M.; Meyer, B.J. Nutritional modulation of cognitive function and mental health. *J. Nutr. Biochem.* **2013**, *24*, 725–743. [CrossRef]
8. Schuchardt, J.P.; Huss, M.; Stauss-Grabo, M.; Hahn, A. Significance of long-chain polyunsaturated fatty acids (PUFAs) for the development and behaviour of children. *Eur. J. Pediatr.* **2010**, *169*, 149–164. [CrossRef]
9. Cooper, R.E.; Tye, C.; Kuntsi, J.; Vassos, E.; Asherson, P. Omega-3 polyunsaturated fatty acid supplementation and cognition: A systematic review and meta-analysis. *J. Psychopharmacol.* **2015**, *29*, 753–763. [CrossRef]
10. Mischoulon, D.; Freeman, M.P. Omega-3 fatty acids in psychiatry. *Psychiatr. Clin. North Am.* **2013**, *36*, 15–23. [CrossRef]
11. Bloch, M.H.; Hannestad, J. Omega-3 fatty acids for the treatment of depression: Systematic review and meta-analysis. *Mol. Psychiatry* **2012**, *17*, 1272–1282. [CrossRef] [PubMed]
12. Politi, P.; Rocchetti, M.; Emanuele, E.; Rondanelli, M.; Barale, F. Randomized Placebo-Controlled Trials of Omega-3 Polyunsaturated Fatty Acids in Psychiatric Disorders: A Review of the Current Literature. *Drug Discov. Technol.* **2013**, *10*, 245–253. [CrossRef]
13. Bloch, M.H.; Qawasmi, A. Omega-3 Fatty Acid Supplementation for the Treatment of Children with Attention-Deficit/Hyperactivity Disorder Symptomatology: Systematic Review and Meta-Analysis. *J. Am. Acad. Child Adolesc. Psychiatry* **2011**, *50*, 991–1000. [CrossRef] [PubMed]
14. Ramalho, R.; Pereira, A.C.; Vicente, F.; Pereira, P. Docosahexaenoic acid supplementation for children with attention deficit hyperactivity disorder: A comprehensive review of the evidence. *Clin. Nutr. ESPEN* **2018**, *25*, 1–7. [CrossRef]
15. Agostoni, C.; Nobile, M.; Ciappolino, V.; Delvecchio, G.; Tesei, A.; Turolo, S.; Crippa, A.; Mazzocchi, A.; Altamura, C.A.; Brambilla, P. The Role of Omega-3 Fatty Acids in Developmental Psychopathology: A Systematic Review on Early Psychosis, Autism and ADHD. *Int. J. Mol. Sci.* **2017**, *18*, 2608. [CrossRef]
16. Lange, K.W.; Hauser, J.; Makulska-Gertruda, E.; Nakamura, Y.; Reissmann, A.; Sakaue, Y.; Takano, T.; Takeuchi, Y. The Role of Nutritional Supplements in the Treatment of ADHD: What the Evidence Says. *Curr. Psychiatry Rep.* **2017**, *19*, 8. [CrossRef]
17. Königs, A.; Kiliaan, A.J. Critical appraisal of omega-3 fatty acids in attention-deficit/hyperactivity disorder treatment. *Neuropsychiatr. Dis. Treat.* **2016**, *12*, 1869–1882.
18. Arnold, L.E. Fish oil is not snake oil. *J. Am. Acad. Child Adolesc. Psychiatry* **2011**, *50*, 969–971. [CrossRef]
19. Nigg, J.T.; Lewis, K.; Edinger, T.; Falk, M. Meta-analysis of attention-deficit/hyperactivity disorder symptoms, restriction diet and synthetic food color additives. *J. Am. Acad. Child Adolesc. Psychiatry* **2012**, *21*, 86–89. [CrossRef]
20. Stevenson, J.; Buitelaar, J.; Cortese, S.; Ferrin, M.; Konofal, E.; Lecendreux, M.; Simonoff, E.; Wong, I.C.; Sonuga-Barke, E. Research review: The role of diet in the treatment of attention-deficit/hyperactivity disorder—An appraisal of the evidence on efficacy and recommendations on the design of future studies. *J. Child Psychol. Psychiatry* **2014**, *55*, 416–427. [CrossRef]

21. European Food Safety Authority (EFSA). Scientific opinion on dietary reference values for fats, including saturated fatty acids, polyunsaturated fatty acids, monounsaturated fatty acids, trans fatty acids, and cholesterol. *EFSA J.* **2010**, *8*, 1461.

22. Morgane, P.J.; Austin-LaFrance, R.; Bronzino, J.; Tonkiss, J.; Díaz-Cintra, S.; Cintra, L.; Kemper, T.; Galler, J.R.; Kemper, T. Prenatal malnutrition and development of the brain. *Neurosci. Biobehav. Rev.* **1993**, *17*, 91–128. [CrossRef]

23. Bourre, J.M.; Dumont, O.; Piciotti, M.; Clément, M.; Chaudière, J.; Bonneil, M.; Nalbone, G.; Lafont, H.; Pascal, G.; Durand, G. Essentiality of omega 3 fatty acids for brain structure and function. *World Rev. Nutr. Diet.* **1991**, *66*, 103–117.

24. Burgess, J.R.; Stevens, L.; Zhang, W.; Peck, L. Long-chain polyunsaturated fatty acids in children with attention-deficit hyperactivity disorder. *Am. J. Clin. Nutr.* **2000**, *71*, 327S–330S. [CrossRef] [PubMed]

25. Pusceddu, M.M.; Kelly, P.; Stanton, C.; Cryan, J.F.; Dinan, T.G. N-3 Polyunsaturated Fatty Acids through the Lifespan: Implication for Psychopathology. *Int. J. Neuropsychopharmacol.* **2016**, *19*. [CrossRef] [PubMed]

26. Kuczmarski, R.J.; Ogden, C.L.; Guo, S.S.; Grummer-Strawn, L.M.; Flegal, K.M.; Mei, Z.; Wei, R.; Curtin, L.R.; Roche, A.F.; Johnson, C.L. CDC Growth Charts for the United States: Methods and development. *Vital Health Stat.* **2002**, *246*, 147–148.

27. Domingo-Salvany, A.; Regidor, E.; Alonso, J.; Alvarez-Dardet, C. Proposal for a social class measure. Working Group of the Spanish Society of Epidemiology and the Spanish Society of Family and Community Medicine. *Aten Primaria* **2000**, *25*, 350. [PubMed]

28. Martin-Moreno, J.M.; Boyle, P.; Gorgojo, L.; Maisonneuve, P.; Fernandez-Rodriguez, J.C.; Salvini, S.; Willett, W.C. Development and Validation of a Food Frequency Questionnaire in Spain. *Int. J. Epidemiol.* **1993**, *22*, 512–519. [CrossRef]

29. Le Moullec, N.; Deheeger, M.; Preziosi, P.; Monteiro, P.; Valeix, P.; Rolland-Cachera, M.F.; Potier De Courcy, G.; Christides, J.P.; Cherouvrier, F.; Galan, P.; et al. Validation du manuel-photos utilisé pour l'enquête alimentaire de l'étude SU. VI. MAX. *Cahiers de Nutrition et de Diététique* **1996**, *31*, 158–164.

30. Moreiras, O.; Carbajal, A.; Cabrera, L.; Cuadrado, C. *Tablas de Composición de Alimentos (Food Composition Tables)*; Ediciones Piramide: Madrid, Spain, 2005.

31. Fernandez-Ballart, J.D.; Piñol, J.L.; Zazpe, I.; Corella, D.; Carrasco, P.; Toledo, E.; Perez-Bauer, M.; Martínez-González, M.Á.; Salas-Salvadó, J.; Martín-Moreno, J.M. Relative validity of a semi-quantitative food-frequency questionnaire in an elderly Mediterranean population of Spain. *Br. J. Nutr.* **2010**, *103*, 1808–1816. [CrossRef]

32. Hepburn, F.N.; Exler, J.; Weihrauch, J.L. Provisional tables on the content of omega-3 fatty acids and other fat components of selected foods. *J. Am. Diet. Assoc.* **1986**, *86*, 788–793.

33. Willett, W.C.; Howe, G.R.; Kushi, L.H. Adjustment for total energy intake in epidemiologic studies. *Am. J. Clin. Nutr.* **1997**, *65*, 1220S–1228S. [CrossRef]

34. Willcutt, E.G. The Prevalence of DSM-IV Attention-Deficit/Hyperactivity Disorder: A Meta-Analytic Review. *Neurotherapeutics* **2012**, *9*, 490–499. [CrossRef] [PubMed]

35. Siró, I.; Kápolna, E.; Lugasi, A. Functional food. Product development, marketing and consumer acceptance. A review. *Appetite* **2008**, *51*, 456–467. [CrossRef] [PubMed]

36. SENC, Sociedad Española de Nutrición Comunitaria. Objetivos nutricionales para la población española. *Rev. Esp. Nutr. Comunitaria* **2011**, *4*, 178–199.

37. Kris-Etherton, P.; Taylor, D.S.; Yu-Poth, S.; Huth, P.; Moriarty, K.; Fishell, V.; Hargrove, R.L.; Zhao, G.; Etherton, T.D. Polyunsaturated fatty acids in the food chain in the United States. *Am. J. Clin. Nutr.* **2000**, *71*, 179S–188S. [CrossRef] [PubMed]

38. Wang, L.J.; Yu, Y.H.; Fu, M.L.; Yeh, W.T.; Hsu, J.L.; Yang, Y.H.; Yang, H.T.; Huang, S.Y.; Wei, I.L.; Chen, W.J.; et al. Dietary Profiles, Nutritional Biochemistry Status, and Attention-Deficit/Hyperactivity Disorder: Path Analysis for a Case-Control Study. *J. Clin. Med.* **2019**, *8*, 709. [CrossRef]

39. FAO/FINUT. Grasas y ácidos grasos en Nutrición Humana. Available online: www.fao.org/3/i1953s/i1953s.pdf (accessed on 10 March 2019).

40. Hawkey, E.; Nigg, J.T. Omega-3 fatty acid and ADHD, blood level analysis and meta-analytic extension of suplementation trials. *Clin. Psychol. Rev.* **2014**, *34*, 496–505. [CrossRef] [PubMed]

41. Guesnet, P.; Tressou, J.; Buaud, B.; Simon, N.; Pasteau, S. Inadequate daily intakes of *n*-3 polyunsaturated fatty acids (PUFA) in the general French population of children (3–10 years), the INCA2 survey. *Eur. J. Nutr.* **2019**, *58*, 895–903. [CrossRef]

42. Marí-Bauset, S.; Llopis-González, A.; Zazpe-García, I.; Marí-Sanchis, A.; Morales-Suárez-Varela, M. Nutritional status of children with autism spectrum disorders (ASDs), a case control study. *J. Autism. Dev. Disord.* **2015**, *45*, 203–212.

43. Bowling, A.B.; Tiemeier, H.W.; Jaddoe, V.W.V.; Barker, E.D.; Jansen, P.W. ADHD symptoms and body composition changes in childhood: A longitudinal study evaluating directionality of associations. *Pediatr. Obes.* **2018**, *13*, 567–575. [CrossRef]

44. Hosomi, R.; Yoshida, M.; Fukunaga, K. Seafood Consumption and Components for Health. *J. Heal. Sci.* **2012**, *4*, 72–86. [CrossRef]

45. Ouellet, V.; Weisnagel, S.J.; Marois, J.; Bergeron, J.; Julien, P.; Gougeon, R.; Tchernof, A.; Holub, B.J.; Jacques, H. Dietary Cod Protein Reduces Plasma C-Reactive Protein in Insulin-Resistant Men and Women. *J. Nutr.* **2008**, *138*, 2386–2391. [CrossRef]

46. Ouellet, V.; Marois, J.; Weisnagel, S.J.; Jacques, H. Dietary cod protein improves insulin sensitivity in insulin-resistant men and women: A randomized controlled trial. *Diabetes Care* **2007**, *30*, 2816–2821. [CrossRef] [PubMed]

47. Jerlich, A.; Fritz, G.; Kharrazi, H.; Hammel, M.; Tschabuschnig, S.; Glatter, O.; Schaur, R. Comparison of HOCl traps with myeloperoxidase inhibitors in prevention of low density lipoprotein oxidation. *Biochim. Biophys. Acta* **2000**, *1481*, 109–118. [CrossRef]

48. Karppi; Rissanen; Nyyssönen; Kaikkonen; Olsson; Voutilainen; Salonen; Karppi, J.; Rissanen, T.H.; Nyyssönen, K.; et al. Effects of Astaxanthin Supplementation on Lipid Peroxidation. *Int. J. Vitam. Nutr.* **2007**, *77*, 3–11. [CrossRef]

49. Yoshida, H.; Yanai, H.; Ito, K.; Tomono, Y.; Koikeda, T.; Tsukahara, H.; Tada, N. Administration of natural astaxanthin increases serum HDL-cholesterol and adiponectin in subjects with mild hyperlipidemia. *Atherosclerosis* **2010**, *209*, 520–523. [CrossRef]

50. Chung, S.Y.; Moriyama, T.; Uezu, E.; Uezu, K.; Hirata, R.; Yohena, N.; Masuda, Y.; Kokubu, T.; Yamamoto, S. Administration of phosphatidylcholine increases brain acetylcholine concentration and improves memory in mice with dementia. *J. Nutr.* **1995**, *125*, 1484–1489. [PubMed]

51. Cortese, S.; Moreira-Maia, C.R.; Fleur, D.S.; Morcillo-Peñalver, C.; Rohde, L.A.; Faraone, S.V. Association Between ADHD and Obesity: A Systematic Review and Meta-Analysis. *Am. J. Psychiatry* **2016**, *173*, 34–43. [CrossRef]

52. Nigg, J.T.; Johnstone, J.M.; Musser, E.D.; Long, H.G.; Willoughby, M.T.; Shannon, J. Attention-deficit/hyperactivity disorder (ADHD) and being overweight/obesity, new data and meta-analysis. *Clin. Psychol. Rev.* **2016**, *43*, 67–79. [CrossRef] [PubMed]

53. Pérez-Bonaventura, I.; Granero, R.; Ezpeleta, L. The relationship between weight status and emotional and behavioral problems in Spanish preschool children. *J. Pediatr. Psychol.* **2015**, *40*, 455–463. [CrossRef]

54. Bass, J.L.; Corwin, M.; Gozal, D.; Moore, C.; Nishida, H.; Parker, S.; Schonwald, A.; Wilker, R.E.; Stehle, S.; Kinane, T.B. The effect of chronic or intermittent hypoxia on cognition in childhood: A review of the evidence. *Pediatrics* **2004**, *114*, 805–816. [CrossRef] [PubMed]

55. Albayrak, Ö.; Pütter, C.; Volckmar, A.L.; Cichon, S.; Hoffmann, P.; Nöthen, M.M.; Jöckel, K.H.; Schreiber, S.; Wichmann, H.E.; Faraone, S.V.; et al. Common obesity risk alleles in childhood attention-deficit/hyperactivity disorder. *Am. J. Med. Genet. B Neuropsychiatr. Genet.* **2013**, *162*, 295–305. [CrossRef] [PubMed]

56. Liu, L.L.; Li, B.M.; Yang, J.; Wang, Y.W. Does dopaminergic reward system contribute to explaining comorbidity obesity and ADHD? *Med. Hypotheses.* **2008**, *70*, 1118–1120. [CrossRef] [PubMed]

57. Ghanadri, Y.; Eisenberg, I.; Ben Neriah, Z.; Agranat-Meged, A.; Kieselstein-Gross, E.; Mitrani-Rosenbaum, S.; Agranat-Meged, A.; Kieselstein-Gross, E.; Mitrani-Rosenbaum, S. Attention deficit hyperactivity disorder in obese melanocortin-4-receptor (MC4R) deficient subjects: A newly described expression of MC4R deficiency. *Am. J. Med. Genet. B Neuropsychiatr. Genet.* **2008**, *147*, 1547–1553.

58. Meyer, B.J.; Mann, N.J.; Lewis, J.L.; Milligan, G.C.; Sinclair, A.J.; Howe, P.R.C. Dietary intakes and food sources of omega-6 and omega-3 polyunsaturated fatty acids. *Lipids* **2003**, *38*, 391–398. [CrossRef]

brain sciences

MDPI

Article

Effects of Transpulmonary Administration of Caffeine on Brain Activity in Healthy Men

Kazutaka Ueda * and Masayuki Nakao

Department of Mechanical Engineering, Graduate School of Engineering, The University of Tokyo, 7-3-1 Hongo, Bunkyo-ku, Tokyo 113-8656, Japan
* Correspondence: ueda@design-i.t.u-tokyo.ac.jp; Tel.: +81-3-5841-0366

Received: 7 July 2019; Accepted: 1 September 2019; Published: 3 September 2019

Abstract: The present study aimed to examine the effect of transpulmonary administration of caffeine on working memory and related brain functions by electroencephalography measurement. The participants performed working memory tasks before and after vaporizer-assisted aspiration with inhalation of caffeinated- and non-caffeinated liquids in the caffeine and sham conditions, respectively. Transpulmonary administration of caffeine tended to increase the rate of correct answers. Moreover, our findings suggest that transpulmonary administration of caffeine increases the theta-band activity in the right prefrontal, central, and temporal areas during the task assigned post-aspiration. Our results may indicate an efficient and fast means of eliciting the stimulatory effects of transpulmonary administration of caffeine.

Keywords: caffeine; transpulmonary administration; working memory; electroencephalography; prefrontal cortex

---

1. Introduction

In recent years, vaporizers have been widely used to gain exhilaration and improve cognitive function. The vaporizer is a device that atomizes liquid by the heat generated from the heating element and performs transpulmonary aspiration of caffeine and herbs, besides nicotine. Oral consumption of caffeine has been reported to improve vigilance [1,2], attention [3,4], memory function [5], and mood [6]. On the contrary, little is known about the effects of transpulmonary administration of caffeine on cognitive function. When administered orally, the peak blood level of caffeine is achieved in 30 to 120 min [7], whereas transpulmonary administration achieves the same in a few seconds [8]. Moreover, caffeine is known to pass through the brain-blood barrier [9]. Based on these facts, transpulmonary administration of caffeine can be expected to have an immediate effect in improving cognitive and related brain functions.

The purpose of this study was to investigate the effects of transpulmonary administration of caffeine on cognitive and related brain functions. We performed electroencephalography (EEG) measurements of participants performing a working memory task before and after the transpulmonary administration of caffeine. We analyzed brain activity in the theta band [10], which is suggested to be related to working memory functions.

2. Materials and Methods

2.1. Participants

Nine healthy male participants (mean age ± standard deviation: 22.8 ± 1.4 years) with normal or corrected-to-normal vision participated in the experiments. None of them had a history of neurological or psychiatric illness. All participants reported being low caffeine consumers (mean consumption = 75 mg/day), and non-smokers. All participants were right-handed, as determined by the Flinders handedness survey (FLANDERS) [11]. This study used a within-subject design to reduce error variance in the physiological measures and has sufficient statistical power to answer the research questions. The study protocol was approved by the Ethics Committee of the Graduate School of Engineering, the University of Tokyo. All participants provided written informed consent prior to their participation in this study.

2.2. Stimuli

We used a commercially available vaporizer, caffeinated liquid (caffeine 1%) for test, and non-caffeinated liquid for the sham condition. Both liquids were transparent and indistinguishable by appearance. Moreover, none of the liquids contained nicotine.

2.3. Experimental Task

The letter 3-back working memory tasks [12] were administered as neurobehavioral probes during EEG measurement. The sequences of the uppercase letters were centrally presented with a stimulus duration of 1000 ms and an interstimulus interval of 1000 ms against a black background using Presentation (Neurobehavioral Systems, Inc., Berkeley, CA, USA). Participants were required to press a button with their right finger immediately if the letter currently presented were the same as the previous three times (Figure 1).

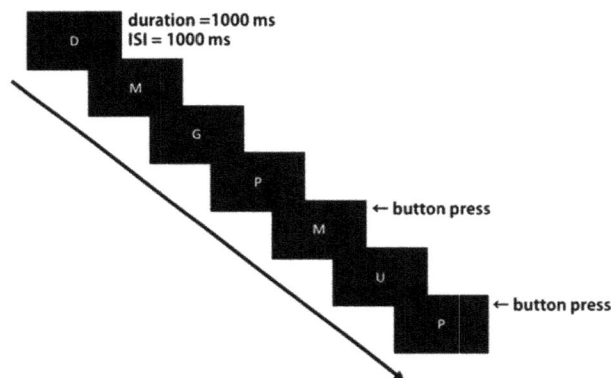

Figure 1. Experimental design of the letter 3-back working memory task.

2.4. Procedures

All participants underwent both caffeine and sham conditions at the same time on two separate days, with an interval of one or more days between the experimental days. The condition order was counterbalanced across participants.

The experiment took place in a shielded room with the participants seated in a comfortable chair, about 90 centimeters from a 29.8″ type display MultiSync LCD-PA302W (NEC Corp, Tokyo, Japan; effective display area of 641 × 401 mm²). The participants were instructed to relax, prevent excessive body or head movements, and to fix their gaze on the middle of the monitor. After explaining the experiment, it was conducted in the following order while with the participants seated on a chair:

(1) Pre-aspiration task: The participants performed the letter 3-back working memory task for 4 min (120 trials).

(2) The participant performed vaporizer aspiration for 2 min.

(3) Post-aspiration task: The participant performed the letter 3-back working memory task for 4 min (120 trials).

The vaporizer aspiration was performed in eight sets with the following steps constituting a set:

(1) Suck steam for 2 s.

(2) Inhale deeply steam for 3 s.

(3) Exhale from his mouth for 6 s.

(4) Rest for 6 s.

The timings of these steps were indicated on the display. The caffeine content in the aspirated vapor in the caffeine condition was about 0.15 mg. Under either of the conditions (caffeine and sham), the participants were required to rate their subjective evaluation concerning the degree of preference and intensity of aroma on a 4-point scale immediately after vaporizer aspiration.

2.5. EEG Recording and Analysis

EEG signals were continuously recorded using the EEG-1200 (Nihon Kohden Corp., Tokyo, Japan) at a sampling rate of 1000 Hz. Nineteen electrodes were positioned according to the international 10−20 system for electrode placement (at the Fp1, Fp2, Fz, F3, F4, F7, F8, Cz, C3, C4, T3, T4, Pz, P3, P4, T5, T6, O1, and O2 sites; Figure 2) [13], using the average of both earlobes as reference, with a time constant of 10 s.

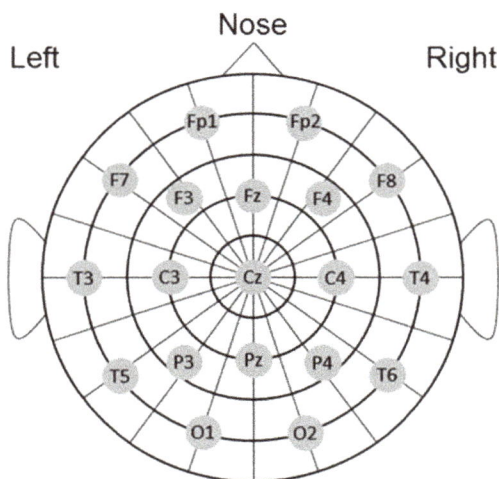

Figure 2. Electroencephalograph electrode positions (Electrode sites of the 10–20 system).

The continuous EEG data were segmented into 4-minute epochs, separately for the pre- and post-aspiration letter 3-back working memory task. The EEG data were exported to EEGLAB14.2b (MATLAB toolbox) [14] for spectral analysis, and were high-pass filtered at 1 Hz using a finite impulse response filter. Electrooculographic artifacts due to blinks or eye movements and electromyographic artifacts were removed using the Automatic Subspace Reconstruction method implemented in the 'clean_rawdata' plugin of EEGLAB [15]. To estimate the average power of the theta band (5–7 Hz), data were processed using the time-frequency algorithm in EEGLAB.

3. Results and Discussion

In order to compare the participants' impressions of the aroma of the vapors in the caffeine and sham conditions, the subjective preference and intensity for aroma were scored (Figures 3 and 4). The paired t-test was performed with the score as the independent variable. There were no significant differences between scores in either of the conditions. We observed that the participants did not feel any difference in the aroma of the vapors under caffeine and sham conditions.

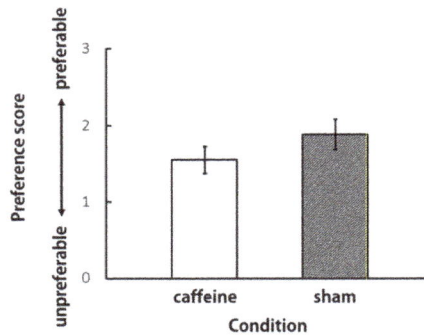

Figure 3. Preference score for the aroma of vapors ($N = 9$).

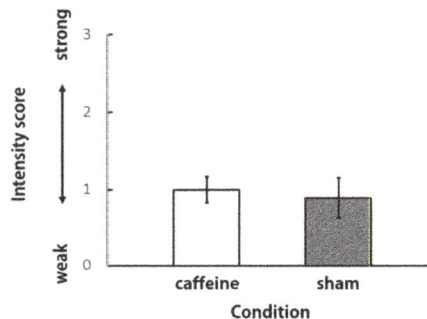

Figure 4. Intensity score for the aroma of vapors ($N = 9$).

In order to compare the behavioral indices in the caffeine and sham conditions, the correct answer rate and the response time in the letter 3-back working memory task were calculated (Figures 5 and 6). In the caffeine condition, the correct answer rate increased post-aspiration, compared to pre-aspiration. Even under the sham condition, the correct answer rate increased post-aspiration; however, this increase was lesser than that under the caffeine condition. There was no difference in the reaction time pre- and post-aspiration under both conditions. In a two-way repeated measures analysis of variance (ANOVA) using treatment (caffeine and sham) and time (pre- and post-aspiration) as the dependent variables, no significant differences were found in the correct answer rate and reaction time.

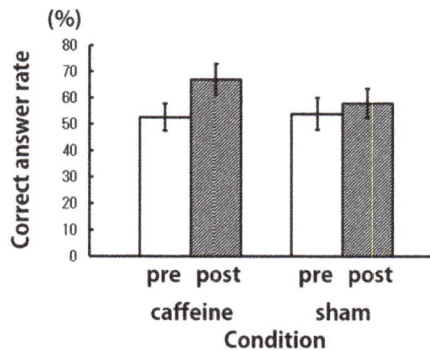

Figure 5. Percentage of correct answer in the letter 3-back working memory task ($N = 9$).

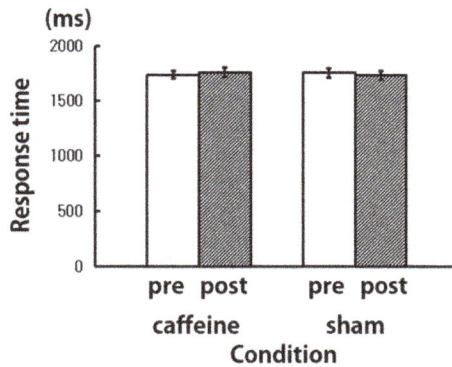

Figure 6. Response time of the letter 3-back working memory task ($N = 9$).

The brain activity in the theta band was calculated for a 4-minute epoch, separately for the pre- and post-aspiration letter 3-back working memory task. A two-way repeated measures ANOVA was performed with treatment (caffeine and sham) and time (pre- and post-aspiration) as the dependent variables, and the log-transformed ($10 \times \log10$ (μV^2)) theta band power (5–7 Hz) as the independent variable. We observed significant interactions for F8, F4, C4, and T4 ($p < 0.05$) (Table 1, Figure 7). Theta-band power differences between post- and pre-aspiration during the letter 3-back working memory task in the caffeine and sham conditions were calculated. The averages of all nine participants in the experiment are presented in Table 1.

Table 1. Theta-band power differences between post- and pre-aspiration during the letter 3-back working memory task in the caffeine and sham conditions ($N = 9$).

	Treatment Condition			
	Caffeine ΔPower (Post–Pre)		Sham ΔPower (Post–Pre)	
EEG Location	Mean	SE	Mean	SE
F8	0.314	0.225	−0.420	0.231
F4	0.164	0.168	−0.416	0.248
C4	0.173	0.171	−0.335	0.189
T4	0.290	0.199	−0.318	0.239

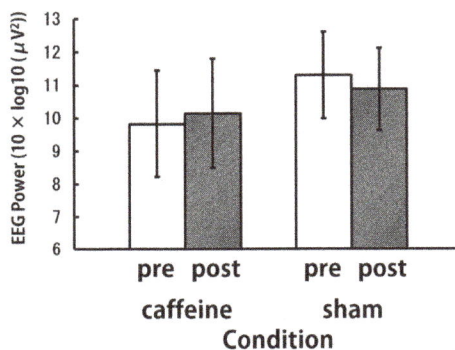

Figure 7. Theta activation at F8 during the letter 3-back working memory task ($N = 9$).

Activity of the right prefrontal cortex (F8 and F4), right central region (C4), and the right temporal region (T4) were enhanced after the aspiration of vapors in the caffeine condition as compared with the sham condition. These results are consistent with previously reported findings, which suggested that the activation of the right frontal area increased during the working memory task after oral administration of caffeine [16,17]. The neuroexcitatory action of caffeine, a non-selective adenosine A1 and A2 receptor antagonist, modulates the activity of the dopamine-rich brain regions of the right hemisphere that are involved in executive and attentional functions required for working memory function [16]. In the previous study, the participants performed the working memory task 20–30 min after oral administration of caffeine, whereas, in this study, the participants performed the task immediately after transpulmonary aspiration of caffeinated vapors. Our findings indicate that transpulmonary administration of caffeine has an immediate effect on the right prefrontal, central, and temporal areas associated with working memory.

With regard to the sham condition, the theta-band activities demonstrated greater deactivation in the post-aspiration task. Previous research has shown the relationship between the sustained effort to focus attention and theta-band activity under working memory load [18]. In the sham condition, it is possible that sustained effort decreased and the theta-band activity decreased with time. On the contrary, the increase in the brain activity for the theta band in the caffeine condition may indicate that caffeine contributes to sustain execution and attention.

In the caffeine condition, the post-aspiration correct answer rate increased more than that of the pre-aspiration. This increase was greater than that observed in the sham condition. However no significant differences were found in the behavioral response. This behavioral effect could be due to the low content of caffeine used in this study rather than the previous study [1–6]. However, as in this study, previous studies of brain function show significant changes in brain activity even without corresponding changes in overt behavior [16,19,20]. The effects of transpulmonary aspiration of caffeine on other brain activity related to resting state, vigilance, attention, and mood are not still clear and need further research.

4. Conclusions

The objective of this study was to investigate the effect of the transpulmonary administration of caffeine on working memory and related brain functions by EEG measurement. The participants performed the letter 3-back working memory tasks before and after vaporizer-assisted aspiration of caffeinated or sham liquid. The transpulmonary administration of caffeine tended to increase the rate of correct answers. Moreover, transpulmonary administration of caffeine was observed to immediately increase the theta-band activity in the right prefrontal, central, and temporal areas during task performance. These results may indicate an efficient and fast means of eliciting the stimulatory effects of transpulmonary administration of caffeine.

Author Contributions: Conceptualization, K.U. and M.N.; methodology, K.U.; software, K.U.; validation, K.U. and M.N.; formal analysis, K.U.; investigation, K.U.; resources, K.U. and M.N.; data curation, K.U.; writing—original draft preparation, K.U.; writing—review and editing, K.U. and M.N.; visualization, K.U.; supervision, M.N.; project administration, M.N.; funding acquisition, K.U. and M.N.

Funding: This research received no external funding.

Conflicts of Interest: The authors declare no conflict of interest.

References

1. Van Dongen, H.P.A.; Price, N.J.; Mullington, J.M.; Szuba, M.P.; Kapoor, S.C.; Dinges, D.F. Caffeine eliminates psychomotor vigilance deficits from sleep inertia. *Sleep* **2001**, *24*, 813–819. [CrossRef] [PubMed]
2. Ramakrishnan, S.; Laxminarayan, S.; Wesensten, N.J.; Kamimori, G.H.; Balkin, T.J.; Reitman, J. Dose-dependent model of caffeine effects on human vigilance during total sleep deprivation. *J. Theor. Biol.* **2014**, *358*, 11–24. [CrossRef] [PubMed]

3. Einother, S.J.L.; Giesbrecht, T. Caffeine as an attention enhancer: Reviewing existing assumptions. *Psychopharmacology* **2013**, *225*, 251–274. [CrossRef] [PubMed]

4. Brunyé, T.T.; Mahoney, C.R.; Lieberman, H.R.; Taylor, H.A. Caffeine modulates attention network function. *Brain Cogn.* **2010**, *72*, 181–188. [CrossRef] [PubMed]

5. Nehlig, A. Is caffeine a cognitive enhancer? *J. Alzheimers Dis.* **2010**, *20*, S85–S94. [CrossRef] [PubMed]

6. Haskell, C.F.; Kennedy, D.O.; Wesnes, K.A.; Scholey, A.B. Cognitive and mood improvements of caffeine in habitual consumers and habitual non-consumers of caffeine. *Psychopharmacology* **2005**, *179*, 813–825. [CrossRef] [PubMed]

7. Blanchard, J.; Sawers, S.J.A. The Absolute Bioavailability of Caffeine in Man. *Eur. J. Clin. Pharmacol.* **1983**, *24*, 93–98. [CrossRef] [PubMed]

8. Zandvliet, A.S.; Huitema, A.D.R.; de Jonge, M.E.; den Hoed, R.; Sparidans, R.W.; Hendriks, V.M.; van den Brink, W.; van Ree, J.M.; Beijnen, J.H. Population pharmacokinetics of caffeine and its metabolites theobromine, paraxanthine and theophylline after inhalation in combination with diacetylmorphine. *Basic Clin. Pharmacol.* **2005**, *96*, 71–79. [CrossRef] [PubMed]

9. Arnaud, M.J. Metabolism of caffeine and other components of coffee. In *Caffeine, Coffee, and Health*; Garattini, S., Ed.; Raven Press: New York, NY, USA, 1993; pp. 43–95.

10. Sauseng, P.; Griesmayr, B.; Freunberger, R.; Klimesch, W. Control mechanisms in working memory: A possible function of EEG theta oscillations. *Neurosci. Biobehav. Rev.* **2010**, *34*, 1015–1022. [CrossRef] [PubMed]

11. Okubo, M.; Suzuki, H.; Nicholls, M.E. A Japanese version of the FLANDERS handedness questionnaire. *Shinrigaku Kenkyu* **2014**, *85*, 474–481. [CrossRef] [PubMed]

12. Owen, A.M.; McMillan, K.M.; Laird, A.R.; Bullmore, E.T. N-back working memory paradigm: A meta-analysis of normative functional neuroimaging. *Hum. Brain Mapp.* **2005**, *25*, 46–59. [CrossRef] [PubMed]

13. Klem, G.H.; Luders, H.O.; Jasper, H.H.; Elger, C. The ten-twenty electrode system of the International Federation. The International Federation of Clinical Neurophysiology. *Electroencephalogr. Clin. Neurophysiol. Suppl.* **1999**, *52*, 3–6. [PubMed]

14. Delorme, A.; Makeig, S. EEGLAB: An open source toolbox for analysis of single-trial EEG dynamics including independent component analysis. *J. Neurosci. Methods* **2004**, *134*, 9–21. [CrossRef] [PubMed]

15. Mullen, T.; Kothe, C.; Chi, Y.M.; Ojeda, A.; Kerth, T.; Makeig, S.; Cauwenberghs, G.; Jung, T.P. Real-Time Modeling and 3D Visualization of Source Dynamics and Connectivity Using Wearable EEG. In Proceedings of the 2013 35th Annual International Conference of the IEEE Engineering in Medicine and Biology Society (EMBC), Osaka, Japan, 3–7 July 2013; pp. 2184–2187.

16. Koppelstaetter, F.; Poeppel, T.D.; Siedentopf, C.M.; Ischebeck, A.; Verius, M.; Haala, I.; Mottaghy, F.M.; Rhomberg, P.; Golaszewski, S.; Gotwald, T. Does caffeine modulate verbal working memory processes? An fMRI study. *NeuroImage* **2008**, *39*, 492–499. [CrossRef] [PubMed]

17. Klaassen, E.B.; de Groot, R.H.M.; Evers, E.A.T.; Snel, J.; Veerman, E.C.I.; Ligtenberg, A.J.M.; Jolles, J.; Veltman, D.J. The effect of caffeine on working memory load-related brain activation in middle-aged males. *Neuropharmacology* **2013**, *64*, 160–167. [CrossRef] [PubMed]

18. Gevins, A.; Smith, M.E.; McEvoy, L.; Yu, D. High-resolution EEG mapping of cortical activation related to working memory: Effects of task difficulty, type of processing, and practice. *Cereb. Cortex* **1997**, *7*, 374–385. [CrossRef] [PubMed]

19. Wilkinson, D.; Halligan, P. Opinion—The relevance of behavioural measures for functional-imaging studies of cognition. *Nat. Rev. Neurosci.* **2004**, *5*, 67–73. [CrossRef] [PubMed]

20. Hershey, T.; Black, K.J.; Hartlein, J.; Braver, T.S.; Barch, D.A.; Carl, J.L.; Perlmutter, J.S. Dopaminergic modulation of response inhibition: An fMRI study. *Cogn. Brain Res.* **2004**, *20*, 438–448. [CrossRef] [PubMed]

brain sciences

MDPI

Perspective

The Counteracting Effects of Exercise on High-Fat Diet-Induced Memory Impairment: A Systematic Review

Paul D. Loprinzi [1], Pamela Ponce [1], Liye Zou [2] and Hong Li [2,3,4,*]

[1] Exercise & Memory Laboratory, Department of Health, Exercise Science and Recreation Management, The University of Mississippi, Oxford, MS 38677, USA; pdloprin@olemiss.edu (P.D.L.); pponce@uthsc.edu (P.P.)

[2] Shenzhen Key Laboratory of Affective and Social Cognitive Science, College of Psychology and Sociology, Shenzhen University, Shenzhen 518060, China; liyezou123@gmail.com

[3] Research Centre of Brain Function and Psychological Science, Shenzhen University, Shenzhen 518060, China

[4] Shenzhen Institute of Neuroscience, Shenzhen University, Shenzhen 518060, China

* Correspondence: lihongszu@szu.edu.cn

Received: 15 May 2019; Accepted: 18 June 2019; Published: 20 June 2019

Abstract: The objective of the present review was to evaluate whether exercise can counteract a potential high-fat diet-induced memory impairment effect. The evaluated databases included: Google Scholar, Sports Discus, Embase/PubMed, Web of Science, and PsychInfo. Studies were included if: (1) an experimental/intervention study was conducted, (2) the experiment/intervention included both a high-fat diet and exercise group, and evaluated whether exercise could counteract the negative effects of a high-fat diet on memory, and (3) evaluated memory function (any type) as the outcome measure. In total, 17 articles met the inclusionary criteria. All 17 studies (conducted in rodents) demonstrated that the high-fat diet protocol impaired memory function and all 17 studies demonstrated a counteracting effect with chronic exercise engagement. Mechanisms of these robust effects are discussed herein.

Keywords: cytokines; hippocampal neurogenesis; inflammation; insulin resistance; obesity

---

1. Introduction

Unlike traditional advice that promotes a low-fat diet [1], recently, high-fat diets (HFDs) are gaining popularity among athletes [2] and the general population [3]. However, HFDs have been shown to impair episodic memory function [4,5]. In humans, episodic memory function refers to the retrospective recall of information from a spatial-temporal context [6]. That is, episodic memory, a contextual-based memory, involves what, where, and when aspects of a memory [7]. In rodents, however, episodic memory is primarily evaluated from a spatial memory task, such as the Morris water maze task or a T-maze task.

As discussed elsewhere [8], a cellular correlate of episodic memory is long-term potentiation (LTP), a form of activity-dependent plasticity that results in enhancement of synaptic transmission [9]. The complementary process of LTP is long-term depression (LTD), in which the efficacy of synaptic transmission is reduced [10]. It is thought that LTP and LTD play an important role in memory as LTP- and LTD-like changes in synaptic strength occur as memories are formed at various sets of brain synapses [11–13]. The adverse episodic memory effects from an HFD may, in part, occur through alterations in processes that influence synaptic transmission and production of plasticity-related proteins [14–16]. For example, research demonstrates that a chronic HFD impairs hippocampal dendritic spine density [17], induces astrocyte alterations [18], reduces expression of the NR2B subunit of NMDA receptors [19], decreases CREB expression [20], and reduces hippocampal BDNF production [21].

Of central interest to this review is whether exercise can counteract HFD-induced memory impairment. Such an effect is plausible for several reasons. We speculate that this counteracting effect may occur from exercise activating some of the neurophysiological pathways that are involved in episodic memory function (e.g., BDNF). Further, we speculate that exercise may counteract HFD-induced memory impairment by, not only activating these pathways, but by also inhibiting the downregulation of these pathways induced by HFD. First, chronic exercise has been shown to enhance episodic memory function [22] and LTP [23]. Chronic exercise may subserve episodic memory function via inducing neurogenesis [23,24] and altering LTP-related receptor (e.g., NMDA) structure and function [25,26].

Couched within the above, HFD may impair episodic memory and exercise has been shown to enhance episodic memory function. Further, exercise has been shown to regulate processes (e.g., LTP) that are impaired with HFD. Thus, the specific research question of this systematic review was to evaluate the extant literature to determine whether exercise can counteract a potential HFD-induced memory impairment effect.

2. Methods

2.1. Computerized Searches

The evaluated databases included: Google Scholar, Sports Discus, Embase/PubMed, Web of Science, and PsychInfo [27]. Articles were retrieved from inception to 22 April 2019. The search terms, including their combinations, were: exercise, physical activity, diet, high-fat, memory, cognition, and cognitive function.

2.2. Study Selection

The computerized searches were performed separately by two authors and comparisons were made to render the number of eligible studies. Consensus was reached from these separate reviews. After conducting the searches, the article titles and abstracts were evaluated to identify applicable articles. Articles meeting the inclusionary criteria were retrieved and evaluated at the full text level.

2.3. Inclusionary Criteria

Studies were included if: (1) an experimental/intervention study was conducted, (2) the experiment/intervention included both an HFD and exercise group, and evaluated whether exercise could counteract the negative effects of an HFD on memory, and (3) evaluated memory function (any type) as the outcome measure.

2.4. Data Extraction of Included Studies

Detailed information from each of the included studies were extracted, including the following information: author, subject characteristics, exercise protocol, diet protocol, temporal assessment of the exercise and diet protocols, memory assessment, whether the diet protocol impaired memory, whether exercise counteracted the diet-induced memory impairment, and evaluated mechanisms of this attenuation effect.

3. Results

3.1. Retrieved Articles

The computerized searches identified 448 articles. Among the 448 articles, 430 were excluded and 18 full text articles were reviewed. Among these 18 articles, 1 was ineligible as it did not meet our study criteria. Thus, in total, 17 articles met the inclusionary criteria and were evaluated herein.

3.2. Article Synthesis

Details on the study characteristics are displayed in Table 1 (extraction table). As shown in Table 1, all studies employed an exercise and diet paradigm in an animal model. The daily exercise protocol ranged from 6 weeks to 23 weeks. Similarly, the diet protocol ranged from 6 weeks to 23 weeks, with the majority of studies implementing an ad libitum diet consisting of 60% fat, 20% carbohydrate, and 20% protein. Across the 17 studies, there was variability on the temporal assessment of the exercise and diet protocols, consisting of either having both protocols occur concurrently or exercise occurring after the high-fat diet (treatment paradigm). Among the 17 studies, 10 implemented a concurrent model, whereas 7 implemented a treatment paradigm. Regarding the memory outcome, the majority of studies utilized a Morris water maze task, with others employing an avoidance task (e.g., passive or step-down) or a maze task (e.g., y-maze task, radial maze task, evaluated plus maze task).

Regarding the effects of HFD on memory, all 17 studies demonstrated that the HFD protocol impaired memory function. Notably, in one study, this impairment effect only occurred among a subgroup of animals (adolescent mice) [28]. Similarly, all 17 studies demonstrated that chronic exercise engagement counteracted HFD-induced memory impairment. Notably, however, one study showed that this attenuation effect only occurred if the chronic exercise protocol occurred during the majority of the HFD period [29].

Table 1. Extraction table of the evaluated studies.

Study	Subjects	Exercise Protocol	Diet Protocol	Temporal Assessment of Exercise and Diet	Memory Assessment	Did High-Fat Diet Impair Memory?	Did Exercise Counteract Diet-Induced Memory Impairment?	Mechanisms
Molteni et al. (2004) [4]	Fisher 344 rats, 2 months old	Free access to running wheel for 2 months.	2 months on high in saturated and monounsaturated fat (primarily from lard plus a small amount of corn oil, approx. 39% energy)	Concurrent	Morris water maze	Yes	Yes	Exercise reversed the decrease in BDNF and its downstream effector, synapsin I (involved in BDNF release). Exercise also increase CREB transcription.
Maesako et al. (2012) [5]	APP transgenic mice overexpressing the familial AD-linked mutation	Enriched environment with access to running wheel; this occurred from weeks 10–20 (i.e., 10 weeks into the high-fat diet).	20 weeks of high-fat diet, involving 60% fat, 20% CHO, and 20% protein	Concurrent	Morris water maze	Yes	Yes	Enriched environment attenuated diet-induced Aβ deposition.
Maesako et al. (2012) [30]	APP transgenic mice overexpressing the familial AD-linked mutation	Voluntary access to running wheel.	20 weeks of high-fat diet, involving 60% fat, 20% CHO, and 20% protein	Concurrent	Morris water maze	Yes	Yes	Exercise attenuated diet-induced Aβ deposition and strengthened the activity of neprilysin, the Aβ-degrading enzyme.
Maesako et al. (2013) [29]	APP transgenic mice overexpressing the familial AD-linked mutation	Voluntary access to running wheel.	20 weeks of high-fat diet, involving 60% fat, 20% CHO, and 20% protein	Concurrent	Morris water maze	Yes	Yes, but only if the exercise occurred throughout the majority of the diet protocol	
Woo et al. (2013) [31]	4-week-old Sprague–Dawley rats	Treadmill exercise for the first 8 weeks, involving a progressive exercise program, ranging from 40 to 60 min/day of exercise.	13-weeks of high-fat diet, involving 45% fat	Concurrent	Morris water maze	Yes	Yes	Upregulation of BDNF and MAPK.

Table 1. *Cont.*

Study	Subjects	Exercise Protocol	Diet Protocol	Temporal Assessment of Exercise and Diet	Memory Assessment	Did High-Fat Diet Impair Memory?	Did Exercise Counteract Diet-Induced Memory Impairment?	Mechanisms
Noble et al. (2014) [32]	7-month-old Naïve rats	Forced treadmill or voluntary wheel access for 7 weeks	16 weeks of high-fat diet	Exercise occurring after high-fat diet (treatment)	Two-way active avoidance test	Yes	Yes	Increased BDNF in CA3.
Cheng et al. (2016) [33]	Twelve-week-old C57BL/6J mice	Treadmill running, 60 min/day, 5 times/week, 15 m/min, for 16 weeks.	16 weeks of high-fat diet ad libitum, involving 60% fat, 20% CHO, and 20% protein	Concurrent	Morris water maze	Yes	Yes	p-CREB, BACE1, IDE, and PSD95 were likely mediators of this effect.
Kang et al. (2016) [34]	Sprague–Dawley rats aged 8 weeks	Treadmill running, 30 min/day, 5 days/week, for 8 weeks.	High fat diet (60% fat) for 20 weeks	Exercise occurring after high-fat diet (treatment)	Passive avoidance task	Yes	Yes	Reduction in pro-inflammatory cytokines (TNF-α, interleukin-1β).
Kim et al. (2016) [35]	Male C57BL/6 mice, 4-weeks old	Treadmill exercise, ranging from 30 to 50 min/day; progressive over a 20-week period.	High-fat diet (60% fat) for 20 weeks ad libitum	Exercise occurring after high-fat diet (treatment)	Y-maze test and radial-8-arm maze test	Yes	Yes	Increased expression of BDNF and TrkB and enhanced cell proliferation.
Klein et al. (2016) [28]	Six-week-old female C57BL/6N mice	Free access to running wheel.	12 weeks of high-fat diet, involving 60% fat, 20% CHO, and 20% protein	Concurrent	Morris water maze	Yes, but only in adolescent	Yes	Modulation of hippocampal neurogenesis.
Park et al. (2017) [36]	Male 4-week-old C57BL/6 mice	Treadmill exercise, 6 days/week, approx. 40 min/day, for 12 weeks.	20 weeks of high-fat diet, involving 60% fat ad libitum	Exercise occurring after high-fat diet (treatment)	Step-down avoidance task	Yes	Yes	Reduced insulin resistance, improved mitochondrial function, and reduced apoptosis in the hippocampus.
Cheng et al. (2018) [37]	Male 3-week-old SHR and normotensive Wistar–Kyoto rats	Swimming exercise for 6 weeks.	6 weeks of low-soybean oil diet	Concurrently	Morris water maze	Yes	Yes	Up-regulation of BDNF and NMDA-r expression.

Table 1. Cont.

Study	Subjects	Exercise Protocol	Diet Protocol	Temporal Assessment of Exercise and Diet	Memory Assessment	Did High-Fat Diet Impair Memory?	Did Exercise Counteract Diet-Induced Memory Impairment?	Mechanisms
Jeong et al. (2018) [38]	Sprague–Dawley rats aged 20 weeks	Treadmill exercise for 8 weeks, 30 min/day, 8 m/min, 5 days/week.	High-fat diet for 20 weeks	Exercise occurring after high-fat diet (treatment)	Water maze and passive avoidance tasks	Yes	Yes	Improved brain insulin signaling, inhibition of obesity-induced NADPH-oxidase activity.
Jeong et al. (2018) [39]	Sprague–Dawley rats aged 8 weeks	Treadmill exercise for 8 weeks, 30 min/day, 5 days/week, progressive intensity.	High fat diet for 20 weeks, including 20% CHO, 60% fat, and 20% protein	Exercise occurring after high-fat diet (treatment)	Passive avoidance task	Yes	Yes	Improved brain insulin signaling (PI3K/AKT/GSK-3β), reduced tau hyperphosphorylation.
Shi et al. (2018) [40]	Male C57BL/6 mice and SIRT3 mice (2-months old)	Exercise started at week 6 and continued for the remaining 6 weeks. Engaged in aerobic intermittent training, 30 min/day, 5 days/week. Intermittent exercise involved 4-min bursts at 80–85% of VO2max, with 2 min active recovery periods.	High-fat diet of 45% kcal fat, 20% kcal protein, and 35% kcal CHO for 12 weeks	Concurrent	Morris water maze	Yes	Yes	SIRT3 upregulation and improvement in antioxidative activity
Han et al. (2019) [41]	Six-week-old C57BL/6 mice	23 weeks of treadmill running, 30 min/day, 5 days/week, at 8 m/min.	23 weeks of high-fat diet ad libitum, involving 60% fat	Concurrent	Morris water maze	Yes	Yes	Reduced number of apoptotic cells and increased BDNF.
Mehta et al. (2019) [42]	Sprague–Dawley male rats	Running wheel access for 6 weeks, 25–30 min/day, 5 days/week.	15 days of high-fat diet (310 gm/kg Lard)	Exercise occurring after high-fat diet (treatment)	Passive avoidance and elevated plus maze	Yes	Yes	Reduction in neuroinflammatory markers (e.g., IL-1β, TNF-α).

4. Discussion

The present review examines whether exercise can counteract HFD-induced memory impairment. Main findings from the present review are twofold: (1) chronic HFD robustly impairs memory function, and (2) chronic exercise engagement, occurring either concurrently or after the diet protocol, robustly counteracted HFD-induced memory impairment. This latter finding occurred among studies that employed various exercise protocols, such as voluntary access to a running wheel or forced exercise on a treadmill. Similarly, across these studies, the exercise protocol varied from 6 to 23 weeks. Further, various spatial-related memory tasks were employed across the evaluated studies. Despite these variations in the exercise protocols and memory tasks, exercise robustly counteracted HFD-induced memory impairment.

A mechanism through which exercise may counteract HFD-induce memory impairment is likely through alterations in processes related to synaptic transmission and production of plasticity-related proteins. As thoroughly addressed elsewhere [43–48], LTP involves several phases, including early-LTP (E-LTP) and late-LTP (L-LTP) [47]. In brief, E-LTP, a protein synthesis-independent process, involves the activation of several kinases (e.g., PKA, CaMKII), which play a critical role in phosphorylating proteins and receptors (e.g., AMPA, NDMA), eventually potentiating synaptic transmission [47]. Endocytosis of such receptors, via, for example, phosphatase activity, may induce LTD [10]. In contrast to E-LTP, L-LTP, a protein synthesis-dependent process, involves gene expression and local protein synthesis via, for example, the TrkB receptor [47]. The following paragraphs link some of these processes to episodic memory function, how HFD impairs these processes, and how exercise influences these processes.

As noted in Table 1 and as shown in Figure 1, potential mechanisms of this exercise-related counteraction effect of HFD-induced memory impairment are multifold. Such effects may include exercise-induced alterations in some of the above-mentioned pathways. For example, activation of the BDNF receptor, TrkB, plays an important role in spatial memory [49]. Specifically, BDNF appears to play a critical role in the consolidation of memories, as previous work demonstrates that continuous intracerebroventricular infusion of antisense BDNF oligonucleotide causes spatial memory deficit [50]. An HFD has been shown to reduce hippocampal BDNF levels and downstream effectors [20], which may lower the neurochemical substrate of the hippocampus that is needed for optimal neuronal function. Exercise may counteract this HFD-induced BDNF reduction and memory impairment via its role in augmenting BDNF levels, via β-hydroxybutyrate alteration [51]. Exercise-induced increases in β-hydroxybutyrate are thought to inhibit histone deacetylases, ultimately facilitating hippocampal BDNF expression [51].

In addition to BDNF, synapsin 1, a neuronal phosphoprotein, plays an important role in regulating neurotransmitter release. A chronic HFD has been shown to lower synapsin 1 levels [20] and reduction of synapsin 1 leads to spatial memory deficit [52,53]. Exercise has been shown to increase synapsin 1 levels [54], which is likely occurring from exercise-induced increases in BDNF (i.e., BDNF may promote the phosphorylation of synapsin 1) [55]. BDNF also plays an important role in hippocampal neurogenesis [56], which may play a causal role in spatial memory. Ablation of adult hippocampal neurogenesis results in impairment of acquiring spatial reference memory [57]. Neurogenesis plays an important role in spatial memory and may, for example, occur via pattern separation mechanisms [58]. A chronic HFD may impair neurogenesis through increases in corticosterone [59], with exercise potentially counteracting this effect via BDNF-mediated hippocampal neurogenesis [60].

In conclusion, this review demonstrated that episodic memory may be impaired with a chronic HFD, yet this effect may be counteracted by chronic engagement in exercise. Future work should consider this model in the context of a preventive paradigm. All of the evaluated studies in this review employed a concurrent or treatment-based model and, thus, it would be worthwhile to evaluate if a period of exercise prior to an HFD protocol can counteract the detrimental effects of an HFD on memory function. Furthermore, future work should also consider evaluating other memory systems (e.g., working memory, episodic memory, procedural memory, prospective memory) to determine whether the observed associations hold true across different memory systems.

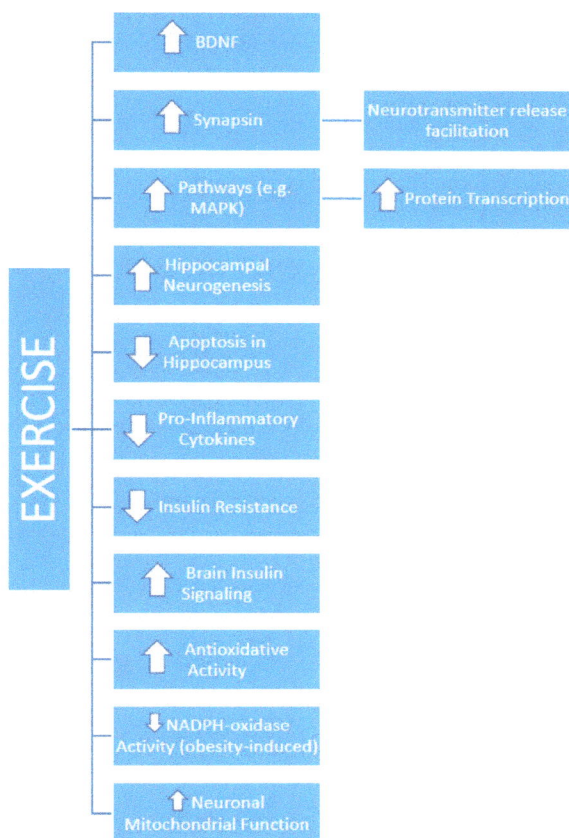

Figure 1. Schematic illustrating the mechanistic role through which exercise may counteract a high-fat diet-induced impairment of memory function.

Funding: This research project is supported by both Guangdong-Government Funding for Scientific Research (2016KZDXM009) and Shenzhen-Government Research Grants Programme in Basic Sciences (JCYJ20150729104249783).

Conflicts of Interest: The authors declare no conflict of interest.

References

1. DeSalvo, K.B.; Olson, R.; Casavale, K.O. Dietary Guidelines for Americans. *JAMA* **2016**, *315*, 1. [CrossRef] [PubMed]
2. Webster, C.C.; Swart, J.; Noakes, T.D.; Smith, J.A. A Carbohydrate Ingestion Intervention in an Elite Athlete Who Follows a Low-Carbohydrate High-Fat Diet. *Int. J. Sports Physiol. Perform.* **2018**, *13*, 957–960. [CrossRef] [PubMed]
3. Noakes, T.D.; Windt, J. Evidence that supports the prescription of low-carbohydrate high-fat diets: A narrative review. *Br. J. Sports Med.* **2017**, *51*, 133–139. [CrossRef]
4. Molteni, R.; Wu, A.; Vaynman, S.; Ying, Z.; Barnard, R.; Gómez-Pinilla, F. Exercise reverses the harmful effects of consumption of a high-fat diet on synaptic and behavioral plasticity associated to the action of brain-derived neurotrophic factor. *Neuroscience* **2004**, *123*, 429–440. [CrossRef]

5. Maesako, M.; Uemura, K.; Kubota, M.; Kuzuya, A.; Sasaki, K.; Asada, M.; Watanabe, K.; Hayashida, N.; Ihara, M.; Ito, H.; et al. Environmental enrichment ameliorated high-fat diet-induced Aβ deposition and memory deficit in APP transgenic mice. *Neurobiol. Aging* **2012**, *33*, 1011.e11–1011.e23. [CrossRef]

6. Tulving, E. *Elements of Episodic Memory*; Oxford University Press: New York, NY, USA, 1983.

7. Mishkin, M.; Suzuki, W.A.; Gadian, D.G.; Vargha-Khadem, F.; Price, T. Hierarchical organization of cognitive memory. *Philos. Trans. R. Soc. B Boil. Sci.* **1997**, *352*, 1461–1467. [CrossRef]

8. Poo, M.-M.; Pignatelli, M.; Ryan, T.J.; Tonegawa, S.; Bonhoeffer, T.; Martin, K.C.; Rudenko, A.; Tsai, L.-H.; Tsien, R.W.; Fishell, G.; et al. What is memory? The present state of the engram. *BMC Boil.* **2016**, *14*, 1133. [CrossRef] [PubMed]

9. Bliss, T.V.P.; Lømo, T. Long-lasting potentiation of synaptic transmission in the dentate area of the anaesthetized rabbit following stimulation of the perforant path. *J. Physiol.* **1973**, *232*, 331–356. [CrossRef]

10. Bol'shakov, V. Mechanisms of long-term synaptic depression in the hippocampus. *Rossiiskii Fiziologicheskii Zhurnal Imeni I.M. Sechenova* **2001**, *87*, 441–447.

11. Doyère, V.; Debiec, J.; Monfils, M.-H.; E Schafe, G.; E LeDoux, J. Synapse-specific reconsolidation of distinct fear memories in the lateral amygdala. *Nat. Neurosci.* **2007**, *10*, 414–416. [CrossRef]

12. Whitlock, J.R.; Heynen, A.J.; Shuler, M.G.; Bear, M.F. Learning induces long-term potentiation in the hippocampus. *Science* **2006**, *313*, 1093–1097. [CrossRef]

13. Gruart, A.; Muñoz, M.D.; Delgado-García, J.M. Involvement of the CA3-CA1 Synapse in the Acquisition of Associative Learning in Behaving Mice. *J. Neurosci.* **2006**, *26*, 1077–1087. [CrossRef] [PubMed]

14. Hao, S.; Dey, A.; Yu, X.; Stranahan, A.M. Dietary obesity reversibly induces synaptic stripping by microglia and impairs hippocampal plasticity. *Brain Behav. Immun.* **2016**, *51*, 230–239. [CrossRef] [PubMed]

15. Valladolid-Acebes, I.; Merino, B.; Principato, A.; Fole, A.; Barbas, C.; Lorenzo, M.P.; Del Olmo, N.; Ruiz-Gayo, M.; Cano, V.; Garcia, A. High-fat diets induce changes in hippocampal glutamate metabolism and neurotransmission. *Am. J. Physiol. Metab.* **2012**, *302*, 396–402. [CrossRef] [PubMed]

16. Karimi, S.A.; Salehi, I.; Komaki, A.; Sarihi, A.; Zarei, M.; Shahidi, S. Effect of high-fat diet and antioxidants on hippocampal long-term potentiation in rats: An in vivo study. *Brain Res.* **2013**, *1539*, 1–6. [CrossRef] [PubMed]

17. Wang, Z.; Fan, J.; Wang, J.; Li, Y.; Xiao, L.; Duan, D.; Wang, Q. Protective effect of lycopene on high-fat diet-induced cognitive impairment in rats. *Neurosci. Lett.* **2016**, *627*, 185–191. [CrossRef]

18. Cano, V.; Valladolid-Acebes, I.; Hernández-Nuño, F.; Merino, B.; Del Olmo, N.; Chowen, J.A.; Ruiz-Gayo, M. Morphological changes in glial fibrillary acidic protein immunopositive astrocytes in the hippocampus of dietary-induced obese mice. *NeuroReport* **2014**, *25*, 819–822. [CrossRef] [PubMed]

19. Page, K.C.; Jones, E.K.; Anday, E.K. Maternal and postweaning high-fat diets disturb hippocampal gene expression, learning, and memory function. *Am. J. Physiol. Integr. Comp. Physiol.* **2014**, *306*, 527–537. [CrossRef] [PubMed]

20. Molteni, R.; Barnard, R.; Ying, Z.; Roberts, C.; Gómez-Pinilla, F. A high-fat, refined sugar diet reduces hippocampal brain-derived neurotrophic factor, neuronal plasticity, and learning. *Neuroscience* **2002**, *112*, 803–814. [CrossRef]

21. Tozuka, Y.; Kumon, M.; Wada, E.; Onodera, M.; Mochizuki, H.; Wada, K. Maternal obesity impairs hippocampal BDNF production and spatial learning performance in young mouse offspring. *Neurochem. Int.* **2010**, *57*, 235–247. [CrossRef]

22. Cassilhas, R.; Lee, K.; Fernandes, J.; Oliveira, M.; Tufik, S.; Meeusen, R.; De Mello, M. Spatial memory is improved by aerobic and resistance exercise through divergent molecular mechanisms. *Neuroscience* **2012**, *202*, 309–317. [CrossRef] [PubMed]

23. Van Praag, H.; Christie, B.R.; Sejnowski, T.J.; Gage, F.H. Running enhances neurogenesis, learning, and long-term potentiation in mice. *Proc. Natl. Acad. Sci. USA* **1999**, *96*, 13427–13431. [CrossRef] [PubMed]

24. Huang, Y.-Q.; Wu, C.; He, X.-F.; Wu, D.; He, X.; Liang, F.-Y.; Dai, G.-Y.; Pei, Z.; Xu, G.-Q.; Lan, Y. Effects of Voluntary Wheel-Running Types on Hippocampal Neurogenesis and Spatial Cognition in Middle-Aged Mice. *Front. Cell. Neurosci.* **2018**, *12*, 177. [CrossRef] [PubMed]

25. Molteni, R.; Ying, Z.; Gomez-Pinilla, F.; Gómez-Pinilla, F. Differential effects of acute and chronic exercise on plasticity-related genes in the rat hippocampus revealed by microarray. *Eur. J. Neurosci.* **2002**, *16*, 1107–1116. [CrossRef] [PubMed]

26. Dietrich, M.O.; Mantese, C.E.; Porciúncula, L.O.; Ghisleni, G.; Vinade, L.; Souza, D.O.; Portela, L.V. Exercise affects glutamate receptors in postsynaptic densities from cortical mice brain. *Brain Res.* **2005**, *1065*, 20–25. [CrossRef] [PubMed]

27. Bramer, W.M.; Rethlefsen, M.L.; Kleijnen, J.; Franco, O.H. Optimal database combinations for literature searches in systematic reviews: A prospective exploratory study. *Syst. Rev.* **2017**, *6*, 245. [CrossRef] [PubMed]

28. Klein, C.; Jonas, W.; Iggena, D.; Empl, L.; Rivalan, M.; Wiedmer, P.; Spranger, J.; Hellweg, R.; Winter, Y.; Steiner, B. Exercise prevents high-fat diet-induced impairment of flexible memory expression in the water maze and modulates adult hippocampal neurogenesis in mice. *Neurobiol. Learn. Mem.* **2016**, *131*, 26–35. [CrossRef]

29. Maesako, M.; Uemura, K.; Iwata, A.; Kubota, M.; Watanabe, K.; Uemura, M.; Noda, Y.; Asada-Utsugi, M.; Kihara, T.; Takahashi, R.; et al. Continuation of Exercise Is Necessary to Inhibit High Fat Diet-Induced β-Amyloid Deposition and Memory Deficit in Amyloid Precursor Protein Transgenic Mice. *PLoS ONE* **2013**, *8*, e72796. [CrossRef]

30. Maesako, M.; Uemura, K.; Kubota, M.; Kuzuya, A.; Sasaki, K.; Hayashida, N.; Asada-Utsugi, M.; Watanabe, K.; Uemura, M.; Kihara, T.; et al. Exercise Is More Effective than Diet Control in Preventing High Fat Diet-induced β-Amyloid Deposition and Memory Deficit in Amyloid Precursor Protein Transgenic Mice. *J. Boil. Chem.* **2012**, *287*, 23024–23033. [CrossRef]

31. Woo, J.; Shin, K.O.; Park, S.Y.; Jang, K.S.; Kang, S. Effects of exercise and diet change on cognition function and synaptic plasticity in high fat diet induced obese rats. *Lipids Heal. Dis.* **2013**, *12*, 144. [CrossRef]

32. Noble, E.E.; Mavanji, V.; Little, M.R.; Billington, C.J.; Kotz, C.M.; Wang, C. Exercise reduces diet-induced cognitive decline and increases hippocampal brain-derived neurotrophic factor in CA3 neurons. *Neurobiol. Learn. Mem.* **2014**, *114*, 40–50. [CrossRef] [PubMed]

33. Cheng, J.; Chen, L.; Han, S.; Qin, L.; Chen, N.; Wan, Z. Treadmill Running and Rutin Reverse High Fat Diet Induced Cognitive Impairment in Diet Induced Obese Mice. *J. Nutr. Health Aging* **2016**, *20*, 503–508. [CrossRef] [PubMed]

34. Kang, E.; Koo, J.; Jang, Y.; Yang, C.; Lee, Y.; Cosio-Lima, L.M.; Cho, J. Neuroprotective Effects of Endurance Exercise against High Fat Diet-Induced Hippocampal Neuroinflammation. *J. Neuroendocr.* **2016**, *28*, 28. [CrossRef] [PubMed]

35. Kim, T.-W.; Choi, H.-H.; Chung, Y.-R. Treadmill exercise alleviates impairment of cognitive function by enhancing hippocampal neuroplasticity in the high-fat diet-induced obese mice. *J. Exerc. Rehabil.* **2016**, *12*, 156–162. [CrossRef] [PubMed]

36. Park, H.S.; Cho, H.S.; Kim, T.W. Physical exercise promotes memory capability by enhancing hippocampal mitochondrial functions and inhibiting apoptosis in obesity-induced insulin resistance by high fat diet. *Metab. Brain Dis.* **2018**, *33*, 283–292. [CrossRef] [PubMed]

37. Cheng, M.; Cong, J.; Wu, Y.; Xie, J.; Wang, S.; Zhao, Y.; Zang, X. Chronic Swimming Exercise Ameliorates Low-Soybean-Oil Diet-Induced Spatial Memory Impairment by Enhancing BDNF-Mediated Synaptic Potentiation in Developing Spontaneously Hypertensive Rats. *Neurochem. Res.* **2018**, *43*, 1047–1057. [CrossRef] [PubMed]

38. Jeong, J.-H.; Koo, J.-H.; Cho, J.-Y.; Kang, E.-B. Neuroprotective effect of treadmill exercise against blunted brain insulin signaling, NADPH oxidase, and Tau hyperphosphorylation in rats fed a high-fat diet. *Brain Res. Bull.* **2018**, *142*, 374–383. [CrossRef]

39. Jeong, J.-H.; Kang, E.-B. Effects of treadmill exercise on PI3K/AKT/GSK-3β pathway and tau protein in high-fat diet-fed rats. *J. Exerc. Nutr. Biochem.* **2018**, *22*, 9–14. [CrossRef] [PubMed]

40. Shi, Z.; Li, C.; Yin, Y.; Yang, Z.; Xue, H.; Mu, N.; Wang, Y.; Liu, M.; Ma, H. Aerobic Interval Training Regulated SIRT3 Attenuates High-Fat-Diet-Associated Cognitive Dysfunction. *BioMed Res. Int.* **2018**, *2018*, 1–8. [CrossRef]

41. Han, T.-K.; Leem, Y.-H.; Kim, H.-S. Treadmill exercise restores high fat diet-induced disturbance of hippocampal neurogenesis through β2-adrenergic receptor-dependent induction of thioredoxin-1 and brain-derived neurotrophic factor. *Brain Res.* **2019**, *1707*, 154–163. [CrossRef]

42. Mehta, B.K.; Singh, K.K.; Banerjee, S. Effect of exercise on type 2 diabetes-associated cognitive impairment in rats. *Int. J. Neurosci.* **2019**, *129*, 252–263. [CrossRef] [PubMed]

43. Blitzer, R.D. Teaching resources. Long-term potentiation: Mechanisms of induction and maintenance. *Sci. STKE* **2005**, *2005*, tr26. [CrossRef] [PubMed]

44. Sweatt, J.D. Toward a Molecular Explanation for Long-Term Potentiation. *Learn. Mem.* **1999**, *6*, 399–416. [CrossRef] [PubMed]

45. Buonarati, O.R.; Hammes, E.A.; Watson, J.F.; Greger, I.H.; Hell, J.W. Mechanisms of postsynaptic localization of AMPA-type glutamate receptors and their regulation during long-term potentiation. *Sci. Signal.* **2019**, *12*, eaar6889. [CrossRef] [PubMed]

46. Sacktor, T.C.; Fenton, A.A. What does LTP tell us about the roles of CaMKII and PKMzeta in memory? *Mol. Brain* **2018**, *11*, 77. [CrossRef] [PubMed]

47. Baltaci, S.B.; Mogulkoc, R.; Baltacim, A.K. Molecular Mechanisms of Early and Late LTP. *Neurochem. Res.* **2019**, *44*, 281–296. [CrossRef] [PubMed]

48. Park, M. AMPA Receptor Trafficking for Postsynaptic Potentiation. *Front. Cell. Neurosci.* **2018**, *12*, 361. [CrossRef] [PubMed]

49. Mizuno, M.; Yamada, K.; He, J.; Nakajima, A.; Nabeshima, T. Involvement of BDNF Receptor TrkB in Spatial Memory Formation. *Learn. Mem.* **2003**, *10*, 108–115. [CrossRef] [PubMed]

50. Mizuno, M.; Yamada, K.; Olariu, A.; Nawa, H.; Nabeshima, T. Involvement of Brain-Derived Neurotrophic Factor in Spatial Memory Formation and Maintenance in a Radial Arm Maze Test in Rats. *J. Neurosci.* **2000**, *20*, 7116–7121. [CrossRef] [PubMed]

51. Sleiman, S.F.; Henry, J.; Al-Haddad, R.; El Hayek, L.; Haidar, E.A.; Stringer, T.; Ulja, D.; Karuppagounder, S.S.; Holson, E.B.; Ratan, R.R.; et al. Exercise promotes the expression of brain derived neurotrophic factor (BDNF) through the action of the ketone body β-hydroxybutyrate. *eLife* **2016**, *5*, 5. [CrossRef] [PubMed]

52. Qiao, S.; Peng, R.; Yan, H.; Gao, Y.; Wang, C.; Wang, S.; Zou, Y.; Xu, X.; Zhao, L.; Dong, J.; et al. Reduction of Phosphorylated Synapsin I (Ser-553) Leads to Spatial Memory Impairment by Attenuating GABA Release after Microwave Exposure in Wistar Rats. *PLoS ONE* **2014**, *9*, e95503. [CrossRef] [PubMed]

53. John, J.P.P.; Sunyer, B.; Höger, H.; Pollak, A.; Lubec, G. Hippocampal synapsin isoform levels are linked to spatial memory enhancement by SGS742. *Hippocampus* **2009**, *19*, 731–738. [CrossRef] [PubMed]

54. Vaynman, S.; Ying, Z.; Gómez-Pinilla, F.; Gómez-Pinilla, F. Exercise induces BDNF and synapsin I to specific hippocampal subfields. *J. Neurosci. Res.* **2004**, *76*, 356–362. [CrossRef]

55. Jovanovic, J.N.; Czernik, A.J.; Fienberg, A.A.; Greengard, P.; Sihra, T.S. Synapsins as mediators of BDNF-enhanced neurotransmitter release. *Nat. Neurosci.* **2000**, *3*, 323–329. [CrossRef] [PubMed]

56. Rossi, C.; Angelucci, A.; Costantin, L.; Braschi, C.; Mazzantini, M.; Babbini, F.; Fabbri, M.E.; Tessarollo, L.; Maffei, L.; Berardi, N.; et al. Brain-derived neurotrophic factor (BDNF) is required for the enhancement of hippocampal neurogenesis following environmental enrichment. *Eur. J. Neurosci.* **2006**, *24*, 1850–1856. [CrossRef] [PubMed]

57. Dupret, D.; Revest, J.-M.; Koehl, M.; Ichas, F.; De Giorgi, F.; Costet, P.; Abrous, D.N.; Piazza, P.V. Spatial Relational Memory Requires Hippocampal Adult Neurogenesis. *PLoS ONE* **2008**, *3*, e1959. [CrossRef]

58. França, T.F.A.; Bitencourt, A.M.; Maximilla, N.R.; Barros, D.M.; Monserrat, J.M. Hippocampal neurogenesis and pattern separation: A meta-analysis of behavioral data. *Hippocampus* **2017**, *27*, 937–950. [CrossRef]

59. Lindqvist, A.; Mohapel, P.; Bouter, B.; Frielingsdorf, H.; Pizzo, D.; Brundin, P.; Erlanson-Albertsson, C.; Erlanson-Albertsson, C.; Erlanson-Albertsson, C. High-fat diet impairs hippocampal neurogenesis in male rats. *Eur. J. Neurol.* **2006**, *13*, 1385–1388. [CrossRef]

60. Lee, J.; Duan, W.; Mattson, M.P. Evidence that brain-derived neurotrophic factor is required for basal neurogenesis and mediates, in part, the enhancement of neurogenesis by dietary restriction in the hippocampus of adult mice. *J. Neurochem.* **2002**, *82*, 1367–1375. [CrossRef]

brain
sciences

MDPI

Review

The Expanding Role of Ketogenic Diets in Adult Neurological Disorders

Tanya J. W. McDonald and Mackenzie C. Cervenka *

Department of Neurology, Johns Hopkins University School of Medicine, 600 North Wolfe Street, Meyer 2-147, Baltimore, MD 21287, USA; twill145@jhmi.edu
* Correspondence: mcerven1@jhmi.edu; Tel.: +1-443-287-0423; Fax: 410-502-2507

Received: 28 June 2018; Accepted: 2 August 2018; Published: 8 August 2018

Abstract: The current review highlights the evidence supporting the use of ketogenic diet therapies in the management of adult epilepsy, adult malignant glioma and Alzheimer's disease. An overview of the scientific literature, both preclinical and clinical, in each area is presented and management strategies for addressing adverse effects and compliance are discussed.

Keywords: modified Atkins diet; epilepsy; glioblastoma multiforme; malignant glioma; Alzheimer's disease

1. Introduction

The ketogenic diet (KD) was formally introduced into practice in the 1920s although the origins of ketogenic medicine may date back to ancient Greece [1]. This high-fat, low-carbohydrate diet induces ketone body production in the liver through fat metabolism with the goal of mimicking a starvation state without depriving the body of necessary calories to sustain growth and development [2,3]. The ketone bodies acetoacetate and β-hydroxybutyrate then enter the bloodstream and are taken up by organs including the brain where they are further metabolized in mitochondria to generate energy for cells within the nervous system. The ketone body acetone, produced by spontaneous decarboxylation of acetoacetate, is rapidly eliminated through the lungs and urine. The classic KD is typically composed of a macronutrient ratio of 4:1 (4 g of fat to every 1 g of protein plus carbohydrates combined), thus shifting the predominant caloric source from carbohydrate to fat. Lower ratios of 3:1, 2:1, or 1:1 (referred to as a modified ketogenic diet) can be used depending on age, individual tolerability, level of ketosis and protein requirements [4]. To increase flexibility and palatability, more 'relaxed' variants have been developed, including the modified Atkins diet (MAD), the low glycemic index treatment (LGIT) and the ketogenic diet combined with medium chain triglyceride oil (MCT). Introduced in 2003, the MAD typically employs a net 10–20 g/day carbohydrate limit which is roughly equivalent to a ratio of 1–2:1 of fat to protein plus carbohydrates [5,6]. The LGIT recommends 40–60 g daily of carbohydrates with the selection of foods with glycemic indices <50 and ~60% of dietary energy derived from fat and 20–30% from protein [7]. The primary goal of this diet, primarily used in children, is not to induce metabolic ketosis and will not be further explored in this review. The MCT variant KD uses medium-chain fatty acids provided in coconut and/or palm kernel oil as a diet supplement and allows for greater carbohydrate and protein intake than even a lower-ratio classic KD [8], which can improve compliance. While there is an extensive literature documenting the use of KDs for weight loss and epilepsy [9,10], these diets have garnered increased interest as potential treatments of other diet-sensitive neurological disorders. The aim of the current review is to describe the evidence, preclinical and clinical, supporting KD use in the management of adult epilepsy, adult malignant gliomas and Alzheimer's disease. Several randomized controlled trials support the use of KDs for the treatment of drug-resistant epilepsy and there is emerging evidence that these diets may also be effective in treating refractory status epilepticus, malignant glioma and Alzheimer's disease in adults.

Brain Sci. **2018**, *8*, 148

2. KDs in the Management of Adult Epilepsy and Refractory Seizures

Despite being first recognized as an effective tool in the treatment of epilepsy in the 1920s [11,12], interest in diet therapy subsequently waned following the introduction of anti-epileptic drugs (AEDs) until the 1990s. Studies and clinical trials emerged demonstrating its efficacy in patients with drug-resistant epilepsy and particular pediatric epilepsy syndromes [11–13]. In the management of drug-resistant epilepsy (seizures resistant to two or more appropriate AEDs), adult patients have a less than 5% chance of seizure freedom with additional drugs added and may not be surgical candidates due to a generalized epilepsy, multifocal nature, or non-resectable seizure focus [14,15]. Seizures that evolve into status epilepticus (prolonged seizure lasting longer than 5 minutes or recurrent seizures without return to baseline between seizures) despite appropriate first- and second-line AEDs are classified as refractory status epilepticus (RSE). If status epilepticus continues or recurs 24 h or more after the initiation of treatment with anesthetic agents to induce burst- or seizure-suppression, patients are diagnosed with super-refractory status epilepticus (SRSE) [16]. Growing preclinical and clinical evidence suggests that KDs can offer seizure reduction and seizure freedom in patients with drug-resistant epilepsy and status epilepticus through a variety of potential mechanisms.

There has been controversy over whether the major ketone bodies produced by the liver are responsible for the anti-seizure effect of the KD primarily due to the clinical observation that blood ketone (i.e., β-hydroxybutyrate) levels inconsistently correlate with seizure control amongst studies [17–21], although findings may relate to diet heterogeneity and methodological differences between studies. In addition, ketone levels at the neuronal or synaptic level may be a more accurate reflection of ketone effects on excitability [22] as opposed to systemic concentrations. As recently reviewed [23], an increasing number of compelling experimental studies highlight pleiotropic anti-seizure and neuroprotective actions of ketones. Such effects include ketone-induced changes in neurotransmitter balance and release as well as changes in neural membrane polarity to dampen the increased neuronal excitability associated with seizures. In rat models of epilepsy, acetoacetate and β-hydroxybutyrate increased the accumulation of γ-aminobutyric acid (GABA) in presynaptic vessels [24]. Ketotic rats, moreover, exhibit lower levels of glutamate in neurons but stable amounts of GABA, suggesting a shift in the total balance of neurotransmitters towards inhibition [25]. Supporting these pre-clinical data, humans maintained on a KD showed increased GABA levels in the cerebrospinal fluid and in brain using magnetic resonance spectroscopy [26,27]. Ketones can slow spontaneous neuronal firing in cultured mouse hippocampal neurons by opening adenosine tri-phosphate (ATP)-sensitive potassium channels [28,29]. Medium chain fatty acids, like decanoic acid, have also exhibited efficacy in *in vitro* and *in vivo* models of seizure activity. Decanoic acid application blocked seizure-like activity in hippocampal slices treated with pentetrazol and increased seizure thresholds in animal models of acute seizures using both the 6 Hz stimulation test (a model of drug-resistant seizures) and the maximal electroshock test (a model of tonic-clonic seizures), potentially through a mechanism involving selective inhibition of AMPA receptors [30–32].

Moreover, there may also be an additional neuroprotective benefit of ketogenic therapies related to improved mitochondrial function due to increased energy reserves combined with decreased production of reactive oxygen species (ROS) [33]. For example, the KD has been shown to stimulate mitochondrial biogenesis, increase cerebral ATP concentrations, and result in lower ROS production in animal models [34,35]. Animal models have similarly demonstrated that the KD may influence seizures associated with the mammalian target of rapamycin (mTOR) pathway, as rats fed a KD showed reduced insulin levels and reduced phosphorylation of Akt and S6, suggesting decreased mTOR activation and increased AMP-activated protein kinase signaling [36,37]. Ketone reduction of oxidative stress may occur via genomic effects, as ketone application in *in vitro* models inhibits histone deactylases (HDACs) resulting in increased transcriptional activity of peroxisome proliferator-activated receptor (PPAR) γ and upregulation of genes including the antioxidants catalase, mitochondrial superoxide dismutase and metallothionein 2 [38,39]. Lastly, there is emerging evidence that ketone bodies exhibit protective anti-inflammatory effects [40]. In animal models, KD treatment reduces microglial activation, expression of pro-inflammatory cytokines and pain and inflammation after thermal nociception [41–43]. Similar experimental work in non-epilepsy models suggested that ketone body anti-inflammatory effects may be mediated by hydroxy-carboxylic acid receptor 2 (HCA2) and/or inhibition of the innate immune sensor NOD-like receptor 3 (NLRP3) inflammasome [23,43,44]. These anti-inflammatory properties may explain the observed benefit of KD in treating patients with SRSE secondary to auto-immune and presumed auto-immune encephalitis such as those with new-onset refractory status epilepticus (NORSE) and febrile infection-related epilepsy syndrome (FIRES) [45].

In contrast to the aforementioned studies highlighting mechanisms mediated largely by ketones, recent preclinical work suggests the anti-seizure properties bestowed by KDs may instead relate to modulation of gut microbiota. The KD has been shown to alter the composition of gut microbiota in mice and ketosis is associated with altered gut microbiota in humans [46–49]. Studies using two mouse models of epilepsy (6 Hz stimulation test and mice harboring a null mutation in the alpha subunit of voltage-gated potassium channel Kv1.1) demonstrate that KD induced changes in gut microbiota, produced by feeding or fecal transplant, are necessary and sufficient to confer seizure protection. The effect appears to be mediated by select microbial interactions that reduce bacterial gamma-glutamylation activity, decrease peripheral gamma-glutamylated-amino acids and elevate bulk hippocampal GABA/glutamate ratios [50]. As rodent studies have shown different taxonomic shifts in response to KD therapy, the gut microbiota induced by KDs will depend on host genetics and baseline metabolic profiles [46,50]. Further research is needed to determine effects of the KD on microbiome profiles in adults with drug-resistant epilepsy and whether particular taxonomic changes in gut microbiota correlate with seizure severity and response to therapy.

A surge of clinical studies since the turn of the century support KD use in the management of chronic epilepsy in adults, with most reporting efficacy defined by the proportion of patients achieving ≥50% seizure reduction (defined as responders). A 2011 review pooled data from seven studies of the classic KD to show that 49% of 206 patients had ≥50% seizure reduction and, of these, 13% were seizure-free [51]. A 2015 meta-analysis reviewing ketogenic dietary treatments in adults from 12 studies of the classic KD, the MAD and the classic KD in combination with MCT found efficacy rates of KDs in drug-resistant epilepsy ranged from 13–70% with a combined efficacy rate of 52% for the classic KD and 34% for the MAD [52]. In the largest observational study of 101 adult patients naïve to diet therapy who subsequently started the MAD, 39% had ≥50% seizure reduction and 22% became seizure-free following 3 months of treatment [53]. Based on intention-to-treat (ITT) data from observational studies to date, the classic KD reduces seizures by ≥50% in 22–70% of patients while the MAD reduces seizures by ≥50% in 12–67% of patients [52,54,55], with some suggestion that dietary intervention may be more effective in patients with generalized rather than focal epilepsy [56,57].

Two randomized controlled trials (RCTs) evaluating MAD efficacy in adults with drug-resistant epilepsy have been reported recently. The first RCT in Iran compared the proportion of patients with focal or generalized epilepsy achieving ≥50% seizure reduction between 34 patients randomized to MAD use for 2 months (of whom 22 completed the study) compared to 32 control patients and found 35.5% (12/34) efficacy in the MAD group (ITT analysis) at 2 months compared to 0% in the control group [58]. These findings are in line with reports from meta-analyses of observational studies using MAD in adults [52]. The second RCT in Norway compared the change in seizure frequency following intervention in patients with drug-resistant (who had tried ≥3 AEDs) focal or multifocal epilepsy randomized to either 12 weeks of MAD (37 patients, of whom 28 received the intervention and 24 completed the study) or their habitual diet (38 patients, of whom 34 received the intervention and 32 completed the study) [59]. While they found no statistically significant difference in seizure frequency nor in 50% responder rate between the two groups following the intervention, a significant reduction in seizure frequency in the diet group compared to controls was observed among patients who completed the study but only for moderate benefit (25–50% seizure reduction). Importantly, compared to the patient population in the Iranian RCT with roughly half generalized and focal epilepsy patients (length of epilepsy 14–17 years on average, 6–9 mean seizures per month and had tried on average 3–4 AEDs), the Norwegian study investigated MAD treatment in adults with solely focal epilepsy who were particularly drug-resistant (length of epilepsy more than 20 years on average, with a median of 15 seizures per month and had tried on average 9–10 AEDs) and did note an improvement in overall seizure severity in the diet group, as measured by the Liverpool Seizure Severity Scale [58,59]. Additional RCTs of larger sample size are warranted to investigate MAD efficacy in different subpopulations of adult epilepsy patients.

Several case reports and case series have also demonstrated the successful use of KD therapy for management of RSE and SRSE [60–64]. For example, a case series of 10 adults with SRSE of median duration 21.5 days treated with a KD (either 4:1 or 3:1 ratio KD) showed successful cessation of status epilepticus in 100% of patients who achieved ketosis (9 out of 10 adults) at a median of 3 days (range 1–31 days) [65]. In the largest phase I/II clinical trial of 15 adult patients treated with a 4:1 ratio KD (14 of whom completed therapy) after a median of 10 days of SRSE, 11 (79% of patients who completed KD therapy, 73% of all patients enrolled) achieved resolution of seizures in a median of 5 days (range 0–10 days) [66]. As both RSE and SRSE carry high rates of morbidity and mortality [67], KDs offer a needed adjunctive strategy for management. KDs have the potential advantages of working rapidly and synergistically with other concurrent treatments; are relatively easy to start, monitor and maintain in the controlled intensive care unit setting with close follow up; do not contribute to hemodynamic instability seen with anesthetic agents and could potentially reduce the need for prolonged use of anesthetic drugs.

3. KDs in the Management of Adult Malignant Gliomas

Malignant gliomas are a highly heterogeneous tumor, refractory to treatment and the most frequently diagnosed primary brain tumor. Glioblastoma multiforme (GBM), the most aggressive type of glioma, carries an exceptionally poor prognosis with a median overall survival duration between 12 and 15 months from time of diagnosis and a 5-year survival rate of less than 5% [68,69]. The current standard of care for treating patients with GBM consists of maximal safe resection, followed by radiotherapy and concurrent chemotherapy with temozolomide [69]. Additional therapeutic strategies include glucocorticoid management of peritumoral edema and anti-angiogenic treatment with bevacicumab (Avastin); however therapeutic progress, particularly in regard to overall survival, remains poor [70]. Emerging research efforts over the past two decades seek to exploit a known cancer hallmark of abnormal energy metabolism in tumor cells named the "Warburg effect" following the discovery of physician, biochemist and Nobel laureate Otto Warburg that tumors exhibit high rates of aerobic glycolysis followed by predominant fermentation of pyruvate to lactate despite sufficient oxygen availability [71,72]. This metabolic phenotype confers several potential advantages to the cancer cell that include (1) more efficient generation of carbon equivalents for macromolecular synthesis; (2) bypassed mitochondrial oxidative metabolism and its concurrent production of reactive oxygen species and (3) acidification of the tumor site to facilitate invasion and progression [73]. As a result of this metabolic alteration, malignant glioma cells critically depend on glucose as the main energy source to survive and sustain their aggressive proliferative properties [74]. Moreover, clinical findings have identified hyperglycemia as a negative predictor of overall survival and a marker of poor prognosis in patients with GBM [75–78]. These findings have prompted nutritional strategies to target glycemic modulation using KDs, caloric restriction, intermittent fasting and combinatorial diet protocols broadly classified as ketogenic metabolic therapy.

Numerous preclinical studies have investigated KDs and/or exogenous supplementation of ketones or ketogenic agents in the treatment of malignant glioma. In the CT-2A malignant mouse astrocytoma model, a calorie-restricted KD decreased plasma glucose, plasma insulin-like growth factor and tumor weight when administered as a stand-alone therapy and elicited potent synergistic anti-cancer effects when administered in combination with glycolytic inhibitor 2-deoxy-D-glucose (2-DG) [79,80]. In the GL-261 malignant glioma model, KD fed mice had reduced peritumoral edema and tumor microvasculature, 20–30% increased median survival time and achieved complete and long-term remission when used concomitantly with radiation therapy [81–83]. A similar synergistic effect was observed between KD and temozolomide in the GL-261 model [84]. Comparable effects of KDs on tumor growth and survival time have also been shown in glioma derived mouse models of metastatic cancer and in patient-derived GBM subcutaneous and orthotopic implantation models [85,86]. These and other studies suggest that KDs induce a metabolic shift in malignant brain tissue towards a pro-apoptotic, anti-angiogenic, anti-invasive and anti-inflammatory state accompanied by a marked reduction in tumor growth in vivo [70] via mechanisms that include:

(1)   Reduction in blood glucose and insulin growth factor-1 levels [79];
(2)   Attenuated insulin activated Akt/mTOR and Ras/mitogen-activated protein kinase (MAPK) signaling pathways [87,88];
(3)   Induction of genes involved in oxidative stress protection and elimination of ROS through histone deactylase inhibition and altered expression of genes related to angiogenesis, vascular remodeling, invasion potential and the hypoxic response [38,82,84];
(4)   Enhanced cytotoxic T cell anti-tumor immunity [89]; and
(5)   Reduced inflammation via ketone body inhibition of the NLRP3 inflammasome and a reduction in other circulating inflammatory markers [43,90].

The first published case report in 2010 of an adult female patient with newly diagnosed GBM treated with a calorie restricted KD concomitant with standard care (radiation plus chemotherapy) following partial surgical resection demonstrated no tumor detection using fluorodeoxyglucose positron emission tomography (FDG-PET) and magnetic resonance imaging (MRI) after two months of treatment. However, after discontinuing diet therapy, tumor recurrence was detected 10 weeks later [91]. Subsequently a retrospective review reported 6 adult patients with newly diagnosed GBM treated with a KD, 4 of whom were alive at a median follow-up of 14 months and demonstrated reduced mean glucose compared to patients on a regular diet but only one patient was without evidence of disease for 12 months at the time of publication [92]. In another case report of 2 adult patients with recurrent GBM treated with a 3:1 calorie-restricted KD, both patients showed evidence of tumor progression by 12 weeks [93]. In the largest pilot trial, of 20 adult patients with recurrent GBM treated with a ≤60 g/day carbohydrate restricted diet, 3 discontinued because of poor tolerability, 3 had stable disease after 6 weeks that lasted 11–13 weeks and 1 had a minor response, with an overall trend towards an increase in progression-free survival in patients with stable ketosis [94]. A more recent case report documented an adult patient with newly diagnosed GBM who continued to experience significant tumor regression 24 months following combined treatment with subtotal resection, calorie restricted KD, hyperbaric oxygen and other targeted metabolic therapies [95]. These early observational, pilot studies and case reports principally provide evidence of feasibility and short-term safety as no serious adverse events were reported. Although they suggest a role for KDs in the management of GBM substantiated by an array of preclinical studies, given study design heterogeneity particularly in regard to diet formulation and calorie restriction, paucity of control groups and differences in endpoints, no conclusive statistical analysis of the clinical impact of KDs on adult GBM patient outcomes can be made. Consequently, a growing scientific interest has led to an increased number of registered clinical trials, including 4 randomized controlled trials (NCT02302235, NCT01754350, NCT01865162 and NCT03075514) of KDs (compared to a standard diet or between two KD types) in the management of adult GBM, 2 of which also include caloric restriction and a primary outcome of overall survival or progression-free survival [70,96].

4. KDs in the Management of Alzheimer's Disease

In Alzheimer's disease (AD), the most common form of progressive dementia, loss of recent memory and cognitive deficits are associated with extracellular deposition of amyloid-β peptide, intracellular tau protein neurofibrillary tangles and hippocampal neuronal death. Theories vary regarding the etiology of the overall disease process but mitochondrial dysfunction and glucose hypometabolism are recognized biochemical hallmarks [97]. Defects in mitochondrial function and a decline in respiratory chain function alter amyloid precursor protein (APP) processing to favor the production of the pathogenic amyloid-β fragment [98]. Reduced uptake and metabolism of glucose have been strongly linked to progressive cognitive degeneration, as neurons starve due to inefficient glycolysis [99]. Moreover, FDG-PET studies find asymptomatic individuals with genetic risk for AD or a positive family history show less prefrontal cortex, posterior cingulate, entorhinal cortex and hippocampal glucose uptake than normal-risk individuals. This reduction is associated with downregulation of the glucose transporter GLUT1 in the brain of individuals with AD [40,100]. Increasing evidence has demonstrated an association between high-glycemic diet and greater cerebral amyloid burden in humans [101] and that increased insulin resistance contributes to the development of sporadic AD [102,103], suggesting diet as a potential modifiable behavior to prevent cerebral amyloid accumulation and reduce AD risk.

Preclinical work supports the role of ketogenic therapies to prevent or ameliorate histological and biochemical changes related to Alzheimer's disease pathology. *In vitro* studies showed attenuation of deleterious amyloid-β induced effects on rat cortical neurons by pre-treatment with coconut oil (containing high concentrations of MCT) or medium chain fatty acids via activation of Akt and extracellular-signal-regulated kinase (ERK) signaling pathways [104]. Similarly, preclinical studies using animal models of dementia demonstrated reduced brain amyloid-β levels, protection from amyloid-β toxicity and better mitochondrial function following administration of the KD, ketones and MCT [105–108]. Importantly, ketone body suppression of mitochondrial amyloid entry has been further shown to improve learning and memory ability in a symptomatic mouse model of AD [109]. In aged rats, a KD administered for 3 weeks improved learning and memory and was associated with increased angiogenesis and capillary density suggesting the KD may support cognition through improved vascular function [110]. In summary, these preclinical observations provide insight into potential mechanisms through which KDs and ketones may influence AD risk and pathology and lay the foundation for subsequent studies in humans.

In the first randomized controlled trial in humans, 20 patients with AD or mild cognitive impairment (MCI) received a single oral dose of either MCT or placebo on separate days and demonstrated expected elevations in serum ketone level following ingestion but only patients without the Apolipoprotein E (APOE) ε4 allele showed enhanced short-term cognitive performance on a brief screening tool measuring cognitive domains that included attention, memory, language and praxis [111]. This study was later replicated with similar improvements in working memory, visual attention and task switching seen in 19 elderly patients without dementia who received the MCT supplement [112]. Another RCT in adults with MCI treated with either a very low (5–10%) or high (50%) carbohydrate diet over 6 weeks showed an improvement in verbal memory performance that correlated with ketone levels in the ketogenic diet group [113]. A 2015 case report suggested that regular ketone monoester ((R)-3-hydroxybutyl (R)-3-hydroxybutyrate) supplementation, rather than a change to habitual diet, produced repeated diurnal elevations in circulating serum β-hydroxybutyrate levels and improved cognitive and daily activity performance over a 20-month period [114]. A single-arm pilot trial in 15 patients with mild-moderate AD using an MCT-supplemented ≥ 1:1 ratio KD for 3 months showed an improved Alzheimer's disease Assessment Scale –cognitive subscale score in 9 out of 10 patients who completed the study and achieved ketosis (as measured by elevated serum β-hydroxybutyrate levels at follow-up) [115]. Three additional studies in patients with MCI or mild-moderate AD using at least 3-month treatment protocols (2 randomized studies of MCT or a ketogenic product compared to placebo for 3–6 months and 1 observational study administering a ketogenic meal over 3 months) reported that the cognitive benefit of ketogenic therapies was greatest in patients who did not have the APOE ε4 allele [116,117] and, in the observational study, was limited to APOE ε4 negative patients with mild AD [118]. A recent study of patients with mild-moderate AD treated with 1 month of MCT supplements demonstrated increased ketone consumption, quantified by brain ^{11}C-acetoacetate PET imaging before and after administration, suggesting ketones from MCT can compensate for the brain glucose deficit observed in AD [119]. The clinical evidence lends support for the use of KDs and/or supplements to improve cognitive outcomes in patients with AD, however results indicate that stage/level of disease progression and APOE ε4 genotype may affect response to dietary treatment. Ongoing registered randomized clinical trials sponsored by Johns Hopkins University (NCT02521818), Wake Forest University (NCT03130036, NCT03472664 and NCT02984540), Université de Sherbrooke (NCT02709356) and the University of British Columbia (NCT02912936) are underway (active, recruiting, or completed) in individuals with subjective memory impairment, mild AD, and/or healthy controls to evaluate the impact of:

(1) 6–18 weeks of a modified Ketogenic-Mediterranean diet compared to a low-fat diet;
(2) 12 weeks of MAD compared to a recommended diet for seniors to achieve a healthy eating index;
(3) 1 month treatment with two different MCT oil emulsions (60–40 oil or C8 oil); or
(4) 10 days, twice a day, supplementation with a lactose-free skim milk drink containing either 10–50 g/day of MCT oil or 10–50 g/day of placebo (high-oleic sunflower oil)

On primary outcomes that include brain acetoacetate/glucose metabolism using PET, AD biomarkers, level of serum ketones, safety and feasibility as well as secondary outcomes that include cognition, function and examining key treatment response variables such as APOE genotype, amyloid positivity and metabolic status that could inform precision medicine approaches to dietary prescription.

5. Management of Adverse Effects and Poor Compliance in Adults

The most commonly reported adverse effects associated with KD use in adults with epilepsy and long-term diet use in children with epilepsy are gastrointestinal effects, weight loss and a transient increase in lipids [120,121]. Similar side effects have been reported in clinical studies of KD use in malignant glioma and AD, although a true assessment of risk in these populations is difficult due to the small number of trials, short duration of follow up and heterogeneity in KD therapy applied [70,122]. The gastrointestinal side effects which include constipation, diarrhea and occasional nausea and vomiting are typically mild, improve with time, can often be managed with diet adjustments with the guidance of a dietitian or nutritionist and infrequently require medical intervention. Smaller meals, increased fiber intake, exercise and increased sodium and fluid intake can often prevent or alleviate these complaints. Weight loss may be an intended positive effect in patients who are overweight but for those who want to maintain or gain weight, adjustments in caloric intake are recommended. This is of particular importance in patients with malignant glioma as the development of cachexia, due to weight loss principally affecting skeletal muscle mass, is associated with decreased cancer therapy tolerance and impaired respiratory function, leading to lower survival rates [123]. Increases in serum lipids have been shown to normalize with continued diet therapy (after 1 year) or return to normal after cessation of diet therapy in adult epilepsy patients [57,124,125]. In addition, very low carbohydrate diets that induce ketosis have been shown to lead to reductions in serum triglycerides, low-density lipoprotein and total cholesterol and increased levels of high-density lipoprotein cholesterol in adults [9]. Other potential side effects can result from vitamin and mineral deficiencies secondary to restricting carbohydrates and prolonged ketonemia, including osteopenia and osteoporosis [3,126,127], although the precise mechanism remains unclear. The standard practice of supplementing a recommended daily allowance of multivitamin and mineral supplements can reduce the risk of such deficiencies.

Diet adherence and compliance remain significant barriers to successful implementation and an adequate assessment of KD efficacy. Common methodologies to assess and document KD adherence in adults beyond patient self-report include frequent measurements of serum β-hydroxybutyrate or urine acetoacetate concentrations during the first few weeks on the diet and/or collection of dietary food records [57,121,128]. As examples, daily urine ketone assessments are traditionally used in adults with epilepsy during MAD initiation until moderate to large levels of ketosis are reached and serum ketone assessments using drops of blood from a finger stick have been used to guide short-term KD therapy in patients with GBM [53,128]. Still, the majority of studies traditionally report adherence based on patient self-report. A combined adherence rate of 45% for all KD types, 38% for the classic KD and 56% for the MAD, has been reported in a review of the epilepsy literature [52]. In the largest observational study of 139 adult epilepsy patients treated with KDs, 48% (67/139) discontinued the diet (39%) or were lost after initial follow up (9%) with approximately half of patients citing difficulty with compliance or restrictiveness as the reason for stopping [53]. The literature of adherence rates in adult GBM and AD is sparse but growing, with the largest GBM study reporting 15% (3/20) drop-out after 2–3 weeks due to subjectively decreased quality of life [94] and a recent 3 month single-arm AD pilot trial of a MCT-supplemented KD reporting 33% (5/15) attrition due to caregiver burden [115]. Often the provision of food recipes and resources to patients and families during initial diet training and subsequent visits can emphasize the variety of food choices and ease of use rather than perceived restrictiveness. Additional methods to improve adherence and compliance, as well as access for patients who live distant from a KD center, include scheduled telephone calls or electronic communication with the supervising dietitian or nutritionist, provision of ketogenic supplements and use of electronic applications like KetoDietCalculator^{TM} (The Charlie Foundation for Ketogenic Therapies, Santa Monica, CA, USA) to prevent drop-out and emphasize progress and success [6,129].

6. Conclusions

Although the neurological conditions discussed in this review-epilepsy, malignant glioma and Alzheimer's disease—have distinct disease processes, each exhibit disrupted energy metabolism, increased oxidative stress and neuro-inflammation. As each of these pathophysiologic factors can be influenced through diet manipulation, it is logical and reasonable that diet could alter the course and outcomes of these and other neurologic disorders that share common pathways. Extensive preclinical work supports the use of KDs and/or ketone bodies to thwart or ameliorate histological and biochemical changes leading to neurologic dysfunction and disease. Demonstrated and hypothesized mechanisms by which ketogenic therapies influence epilepsy, malignant glioma and Alzheimer's disease include metabolic regulation, neurotransmission modulation, reduced oxidative stress and anti-inflammatory and genomic effects that were highlighted in this review and summarized in Table 1. In some instances, an understanding of the mechanisms by which the KD and ketones exert their effects has led to novel therapeutic targets and work to develop new pharmaceutical drugs [130]. For many disorders, the clinical literature is still growing and limited by the conditions that make dietary interventions difficult to evaluate. For example, evaluation of whole diet changes cannot be performed under blinded circumstances as the participant will be aware of the diet changes made and/or the content of their meals. Other methodological constraints relate to limitations in inter-study comparison due to the heterogeneity of diet intervention used and reduced statistical power to detect significant effects when baseline levels of nutrient intake and individual variability are appropriately controlled. There are also challenges in monitoring diet compliance in the ambulatory setting as adherence to a prescribed diet can be more difficult to achieve than with a traditional pharmaceutical intervention. However, dietary interventions have the advantage of being non-invasive, relatively low risk and generally without serious adverse effects in the appropriate clinical context and may be particularly useful as an adjunctive therapy that synergizes with other pharmacologic and non-pharmacologic approaches. The scientific evidence collected from clinical studies in humans to date has supported KD therapy use in adult epilepsy, adult malignant glioma and Alzheimer's disease, although overall assessment of efficacy remains limited due to study heterogeneity and indications that particular patient subpopulations may achieve disparate levels of benefit. Further clinical investigation using more standardized KD protocols and in patient subpopulations is warranted.

Brain Sci. **2018**, *8*, 148

Table 1. Hypothesized mechanisms through which ketogenic therapies influence neurological disease.

Ketogenic Mechanisms	Epilepsy	Malignant Glioma	Alzheimer's Disease
Metabolic Regulation			
↓Glucose uptake & glycolysis	+	+	
↓Insulin, IGF1 signaling		+	+
↑Ketones/ketone metabolism	+		+
Altered gut microbiota	+		
Neurotransmission			
Altered balance of excitatory/inhibitory neurotransmitters	+		
Inhibition of AMPA receptors	+		
↓mTOR activation & signaling	+	+	
Modulation of ATP-sensitive potassium channels	+		
Oxidative Stress			
↓Production of reactive oxygen species	+	+	
↑Mitochondrial biogenesis/function	+		+
Inflammation/Neuroprotection			
↓Inflammatory cytokines	+	+	
NLRP3 inflammasome inhibition	+	+	
↑cytotoxic T cell function		+	
↓peritumoral edema		+	
↓amyloid-β levels			+
Genomic Effects			
Inhibition of HDACs	+	+	
↑PPARγ	+		
↓Expression of angiogenic factors in tumor cells		+	

AMPA—α-amino-3-hydroxyl-5-methyl-4-isoxazolepropionic acid; IGF1—insulin-like growth factor 1; HDACs—histone deacetylases; mTOR—mammalian target of rapamycin; NLRP3—NOD-like receptor protein 3; PPAR—peroxisome proliferator-activated receptor. ↓—decreased; ↑—increased; +—mechanism shown in *in vitro* or *in vivo* studies.

Funding: This research received no external funding.

Acknowledgments: We would like to acknowledge the multidisciplinary team at the Johns Hopkins Adult Epilepsy Diet Center-Eric Kossoff, Rebecca Fisher, Joanne Barnett, Bobbie Henry-Barron and Diane Vizthum-as well as our patients and their families.

Conflicts of Interest: Mackenzie C. Cervenka has received grant support from Nutricia North America, Vitaflo, Army Research Laboratory, The William and Ella Owens Medical Research Foundation and BrightFocus Foundation. She receives speaking honoraria from LivaNova, Nutricia North America and the Glut1 Deficiency Foundation and performs consulting with Nutricia North America and Sage Therapeutics and Royalties from Demos Health.

References

1. Hippocrates. On the Sacred Disease. Available online: http://classics.mit.edu/Hippocrates/sacred.html (accessed on 16 May 2017).
2. McNally, M.A.; Hartman, A.L. Ketone bodies in epilepsy. *J. Neurochem.* **2012**, *121*, 28–35. [CrossRef] [PubMed]
3. Cervenka, M.C.; Kossoff, E.H. Dietary treatment of intractable epilepsy. *Continuum (Minneap Minn.)* **2013**, *19*, 756–766. [CrossRef] [PubMed]
4. Zupec-Kania, B.A; Spellman, E. An overview of the ketogenic diet for pediatric epilepsy. *Nutr. Clin. Pract.* **1998**, *23*, 589–596. [CrossRef] [PubMed]
5. Kossoff, E.H.; Rowley, H.; Sinha, S.R.; Vining, E.P.G. A prospective study of the modified Atkins diet for intractable epilepsy in adults. *Epilepsia* **2008**, *49*, 316–319. [CrossRef] [PubMed]
6. Cervenka, M.C.; Terao, N.N.; Bosarge, J.L.; Henry, B.J.; Klees, A.A.; Morrison, P.F.; Kossoff, E.H. E-mail management of the modified Atkins diet for adults with epilepsy is feasible and effective. *Epilepsia* **2012**, *53*, 728–732. [CrossRef] [PubMed]

7. Muzykewicz, D.A.; Lyczkowski, D.A.; Memon, N.; Conant, K.D.; Pfeifer, H.H.; Thiele, E.A. Efficacy, safety and tolerability of the low glycemic index treatment in pediatric epilepsy. *Epilepsia* **2009**, *50*, 1118–1126. [CrossRef] [PubMed]

8. Neal, E.G.; Cross, J.H. Efficacy of dietary treatments for epilepsy. *J. Hum. Nutr. Diet.* **2010**, *23*, 113–119. [CrossRef] [PubMed]

9. Paoli, A.; Rubini, A.; Volek, J.S.; Grimaldi, K.A. Beyond weight loss: A review of the therapeutic uses of very-low-carbohydrate (ketogenic) diets. *Eur. J. Clin. Nutr.* **2013**, *67*, 789–796. [CrossRef] [PubMed]

10. McDonald, T.J.W.; Cervenka, M.C. Ketogenic diets for adults with highly refractory epilepsy. *Epilepsy Curr.* **2017**, *17*. [CrossRef] [PubMed]

11. Barborka, C.J. Ketogenic diet treatment of epilepsy in adults. *JAMA* **1928**, *9*, 73–78. [CrossRef]

12. Barborka, C.J. Epilepsy in adults: Results of treatment by ketogenic diet in one hundred cases. *Arch. Neurol. Psych.* **1930**, *23*, 904–914. [CrossRef]

13. Martin, K.; Jackson, C.F.; Levy, R.G.; Cooper, P.N. Ketogenic diet and other dietary treatments for epilepsy. *Cochrane Database Syst. Rev.* **2016**. [CrossRef] [PubMed]

14. Brodie, M.J.; Barry, S.J.E.; Bamagous, G.A.; Norrie, J.D.; Kwan, P. Patterns of treatment response in newly diagnosed epilepsy. *Neurology* **2012**, *78*, 1548–1554. [CrossRef] [PubMed]

15. Chen, Z.; Brodie, M.J.; Liew, D.; Kwan, P. Treatment outcomes in patients with newly diagnosed epilepsy treated with established and new antiepileptic drugs a 30-year longitudinal cohort study. *JAMA Neurol.* **2018**, *75*, 279–286. [CrossRef] [PubMed]

16. Hocker, S.E.; Britton, J.W.; Mandrekar, J.N.; Wijdicks, E.F.M.; Rabinstein, A.A. Predictors of outcome in refractory status epilepticus. *JAMA Neurol.* **2013**, *70*, 72–77. [CrossRef] [PubMed]

17. Gilbert, D.L.; Pyzik, P.L.; Freeman, J.M. The ketogenic diet: Seizure control correlates better with serum beta-hydroxybutyrate than with urine ketones. *J. Child Neurol.* **2000**, *15*, 787–790. [CrossRef] [PubMed]

18. Van Delft, R.; Lambrechts, D.; Verschuure, P.; Hulsman, J.; Majoie, M. Blood beta-hydroxybutyrate correlates better with seizure reduction due to ketogenic diet than do ketones in the urine. *Seizure* **2010**, *19*, 36–39. [CrossRef] [PubMed]

19. Kossoff, E.H.; Zupec-Kania, B.A.; Amark, P.E.; Ballaban-Gil, K.R.; Christina Bergqvist, A.G.; Blackford, R.; Buchhalter, J.R.; Caraballo, R.H.; Helen Cross, J.; Dahlin, M.G.; et al. Optimal clinical management of children receiving the ketogenic diet: Recommendations of the International Ketogenic Diet Study Group. *Epilepsia* **2009**, *50*, 304–317. [CrossRef] [PubMed]

20. Kossoff, E.H.; Rho, J.M. Ketogenic diets: Evidence for short- and long-term efficacy. *Neurotherapeutics* **2009**, *6*, 406–414. [CrossRef] [PubMed]

21. Buchhalter, J.R.; D'Alfonso, S.; Connolly, M.; Fung, E.; Micholas, A.; Sinasac, D.; Singer, R.; Smith, J.; Singh, N.; Rho, J.M. The relationship between d-beta-hydroxybutyrate blood concentrations and seizure control in children treated with the ketogenic diet for medically intractable epilepsy. *Epilepsia Open* **2017**, *2*, 317–321. [CrossRef] [PubMed]

22. Stafstrom, C.E. Dietary Approaches to Epilepsy Treatment: Old and New Options on the Menu. *Epilepsy Curr.* **2004**, *4*, 215–222. [CrossRef] [PubMed]

23. Simeone, T.A.; Simeone, K.A.; Stafstrom, C.E.; Rho, J.M. Do ketone bodies mediate the anti-seizure effects of the ketogenic diet? *Neuropharmacology* **2018**, *133*, 233–241. [CrossRef] [PubMed]

24. Erecińska, M.; Nelson, D.; Daikhin, Y.; Yudkoff, M. Regulation of GABA level in rat brain synaptosomes: Fluxes through enzymes of the GABA shunt and effects of glutamate, calcium and ketone bodies. *J. Neurochem.* **1996**, *67*, 2325–2334. [CrossRef] [PubMed]

25. Melø, T.M.; Nehlig, A.; Sonnewald, U. Neuronal-glial interactions in rats fed a ketogenic diet. *Neurochem. Int.* **2006**, *48*, 498–507. [CrossRef] [PubMed]

26. Wang, Z.J.; Bergqvist, C.; Hunter, J.V.; Jin, D.; Wang, D.J.; Wehrli, S.; Zimmerman, R.A. In vivo measurement of brain metabolites using two-dimensional double-quantum MR spectroscopy-Exploration of GABA levels in a ketogenic diet. *Magn. Reson. Med.* **2003**, *49*, 615–619. [CrossRef] [PubMed]

27. Dahlin, M.; Elfving, Å.; Ungerstedt, U.; Åmark, P. The ketogenic diet influences the levels of excitatory and inhibitory amino acids in the CSF in children with refractory epilepsy. *Epilepsy Res.* **2005**, *64*, 115–125. [CrossRef] [PubMed]

28. Tanner, G.R.; Lutas, A.; Martinez-Francois, J.R.; Yellen, G. Single KATP Channel Opening in Response to Action Potential Firing in Mouse Dentate Granule Neurons. *J. Neurosci.* **2011**, *31*, 8689–8696. [CrossRef] [PubMed]

29. Ma, W.; Berg, J.; Yellen, G. Ketogenic Diet Metabolites Reduce Firing in Central Neurons by Opening KATP Channels. *J. Neurosci.* **2007**, *27*, 3618–3625. [CrossRef] [PubMed]

30. Tan, K.N.; Carrasco-Pozo, C.; McDonald, T.S.; Puchowicz, M.; Borges, K. Tridecanoin is anticonvulsant, antioxidant and improves mitochondrial function. *J. Cereb. Blood Flow Metab.* **2017**, *37*, 2035–2048. [CrossRef] [PubMed]

31. Wlaź, P.; Socała, K.; Nieoczym, D.; Zarnowski, T.; Zarnowska, I.; Czuczwar, S.J.; Gasior, M. Acute anticonvulsant effects of capric acid in seizure tests in mice. *Prog. Neuropsychopharmacol. Biol. Psychiatry* **2015**, *57*, 110–116. [CrossRef] [PubMed]

32. Chang, P.; Augustin, K.; Boddum, K.; Williams, S.; Sun, M.; Terschak, J.A.; Hardege, J.D.; Chen, P.E.; Walker, M.C.; Williams, R.S.B. Seizure control by decanoic acid through direct AMPA receptor inhibition. *Brain* **2016**, *139*, 431–443. [CrossRef] [PubMed]

33. Maalouf, M.; Rho, J.M.; Mattson, M.P. The neuroprotective properties of calorie restriction, the ketogenic diet and ketone bodies. *Brain Res. Rev.* **2009**, *59*, 293–315. [CrossRef] [PubMed]

34. Sullivan, P.G.; Rippy, N.A.; Dorenbos, K.; Concepcion, R.C.; Agarwal, A.K.; Rho, J.M. The Ketogenic Diet Increases Mitochondrial Uncoupling Protein Levels and Activity. *Ann. Neurol.* **2004**, *55*, 576–580. [CrossRef] [PubMed]

35. Bough, K.J.; Wetherington, J.; Hassel, B.; Pare, J.F.; Gawryluk, J.W.; Greene, J.G.; Shaw, R.; Smith, Y.; Geiger, J.D.; Dingledine, R.J. Mitochondrial Biogenesis in the Anticonvulsant Mechanism of the Ketogenic Diet. *Ann. Neurol.* **2006**, *60*, 223–235. [CrossRef] [PubMed]

36. Yamada, K.A. Calorie restriction and glucose regulation. *Epilepsia* **2008**, *49*, 94–96. [CrossRef] [PubMed]

37. McDaniel, S.S.; Rensing, N.R.; Thio, L.L.; Yamada, K.A.; Wong, M. The ketogenic diet inhibits the mammalian target of rapamycin (mTOR) pathway. *Epilepsia* **2011**, *52*, 7–11. [CrossRef] [PubMed]

38. Shimazu, T.; Hirschey, M.; Newman, J.; He, W.; Shirakawa, K.; Moan, N.L.; Grueter, C.A.; Lim, H.; Saunders, L.R.; Stevens, R.D.; et al. Suppression of Oxidative Stress by β-Hydroxybutyrate, an Endogenous Histone Deacetylase Inhibitor. *Science* **2013**, *339*, 211–214. [CrossRef] [PubMed]

39. Jeong, E.A.; Jeon, B.T.; Shin, H.J.; Kim, N.; Lee, D.H.; Kim, H.J.; Kang, S.S.; Cho, G.J.; Choi, W.S.; Roh, G.S. Ketogenic diet-induced peroxisome proliferator-activated receptor-γ activation decreases neuroinflammation in the mouse hippocampus after kainic acid-induced seizures. *Exp. Neurol.* **2011**, *232*, 195–202. [CrossRef] [PubMed]

40. Koppel, S.J.; Swerdlow, R.H. Neurochemistry International Neuroketotherapeutics: A modern review of a century-old therapy. *Neurochem. Int.* **2017**. [CrossRef]

41. Ruskin, D.N.; Kawamura, M.; Masino, S.A. Reduced Pain and Inflammation in Juvenile and Adult Rats Fed a Ketogenic Diet. *PLoS ONE* **2009**, *4*, 1–6. [CrossRef] [PubMed]

42. Yang, X.; Cheng, B. Neuroprotective and Anti-inflammatory Activities of Ketogenic Diet on MPTP-induced Neurotoxicity. *J. Mol. Neurosci.* **2010**, 145–153. [CrossRef] [PubMed]

43. Youm, Y.; Nguyen, K.Y.; Grant, R.W.; Goldberg, E.L.; Bodogai, M.; Kim, D.; Agostino, D.D.; Planavsky, N.; Lupfer, C.; Kanneganti, T.D.; et al. The ketone metabolite β-hydroxybutyrate blocks NLRP3 inflammasome–mediated inflammatory disease. *Nat. Med.* **2015**, *21*, 263–269. [CrossRef] [PubMed]

44. Rahman, M.; Muhammad, S.; Khan, M.A.; Chen, H.; Ridder, D.A.; Müller-Fielitz, H.; Pokorná, B.; Vollbrandt, T.; Stölting, I.; Nadrowitz, R.; et al. The β-hydroxybutyrate receptor HCA 2 activates a neuroprotective subset of macrophages. *Nat. Commun.* **2014**, *5*, 1–11. [CrossRef] [PubMed]

45. Gaspard, N.; Hirsch, L.J.; Sculier, C.; Loddenkemper, T.; Van Baalen, A.; Lancrenon, J.; Emmery, M.; Specchio, N.; Farias-Moeller, R.; Wong, N.; et al. New-onset refractory status epilepticus (NORSE) and febrile infection–related epilepsy syndrome (FIRES): State of the art and perspectives. *Epilepsia* **2018**, *59*, 745–752. [CrossRef] [PubMed]

46. Klein, M.S.; Newell, C.; Bomhof, M.R.; Reimer, R.A.; Hittel, D.S.; Rho, J.M.; Vogel, H.J.; Shearer, J. Metabolomic Modeling to Monitor Host Responsiveness to Gut Microbiota Manipulation in the BTBR [T+tf/j] Mouse. *J. Proteome Res.* **2016**, *15*, 1143–1150. [CrossRef] [PubMed]

47. Newell, C.; Bomhof, M.R.; Reimer, R.A.; Hittel, D.S.; Rho, J.M.; Shearer, J. Ketogenic diet modifies the gut microbiota in a murine model of autism spectrum disorder. *Mol. Autism.* **2016**, *7*, 1–6. [CrossRef] [PubMed]

48. Duncan, S.H.; Lobley, G.E.; Holtrop, G.; Ince, J.; Johnstone, A.M.; Louis, P.; Flint, H.J. Human colonic microbiota associated with diet, obesity and weight loss. *Int. J. Obes.* **2008**, *32*, 1720–1724. [CrossRef] [PubMed]

49. David, L.A.; Maurice, C.F.; Carmody, R.N.; Gootenberg, D.B.; Button, J.E.; Wolfe, B.E.; Ling, A.V.; Devlin, A.S.; Varma, Y.; Fischbach, M.A.; et al. Diet rapidly and reproducibly alters the human gut microbiome. *Nature* **2014**, *505*, 559–563. [CrossRef] [PubMed]

50. Olson, C.A.; Vuong, H.E.; Yano, J.M.; Liang, QY.; Nusbaum, D.J.; Hsiao, E.Y. The Gut Microbiota Mediates the Anti-Seizure Effects of the Ketogenic Diet. *Cell* **2018**, *173*, 1728–1741. [CrossRef] [PubMed]

51. Payne, N.E.; Cross, J.H.; Sander, J.W.; Sisodiya, S.M. The ketogenic and related diets in adolescents and adults-A review. *Epilepsia* **2011**, *52*, 1941–1948. [CrossRef] [PubMed]

52. Ye, F.; Li, X.J.; Jiang, W.L.; Sun, H.B.; Liu, J. Efficacy of and patient compliance with a ketogenic diet in adults with intractable epilepsy: A meta-analysis. *J. Clin. Neurol.* **2015**, *11*, 26–31. [CrossRef] [PubMed]

53. Cervenka, M.C.; Henry, B.J.; Felton, E.A.; Patton, K.; Kossoff, E.H. Establishing an Adult Epilepsy Diet Center: Experience, efficacy and challenges. *Epilepsy Behav.* **2016**, *58*, 61–68. [CrossRef] [PubMed]

54. Williams, T.; Cervenka, M.C. The role for ketogenic diets in epilepsy and status epilepticus in adults. *Clin. Neurophysiol. Pract.* **2017**, *2*, 154–160. [CrossRef]

55. Liu, H.; Yang, Y.; Wang, Y.; Tang, H.; Zhang, F.; Zhang, Y.; Zhao, Y. Ketogenic diet for treatment of intractable epilepsy in adults: A meta-analysis of observational studies. *Epilepsia Open* **2018**, *3*, 9–17. [CrossRef] [PubMed]

56. Kverneland, M.; Selmer, K.K.; Nakken, K.O.; Iversen, P.O.; Taubøll, E. A prospective study of the modified Atkins diet for adults with idiopathic generalized epilepsy. *Epilepsy Behav.* **2015**, *53*, 197–201. [CrossRef] [PubMed]

57. Klein, P.; Janousek, J.; Barber, A.; Weissberger, R. Ketogenic diet treatment in adults with refractory epilepsy. *Epilepsy Behav.* **2010**, *19*, 575–579. [CrossRef] [PubMed]

58. Zare, M.; Okhovat, A.A.; Esmaillzadeh, A.; Mehvari, J.; Najafi, M.R.; Saadatnia, M. Modified atkins diet in adult patients with refractory epilepsy: A controlled randomized clinical trial. *Iran. J. Neurol.* **2017**, *16*, 72–77. [CrossRef]

59. Kverneland, M.; Molteberg, E.; Iversen, P.O.; Veierød, M.B.; Taubøll, E.; Selmer, K.K.; Nakken, K.O. Effect of modified Atkins diet in adults with drug-resistant focal epilepsy: A randomized clinical trial. *Epilepsia* **2018**, 1–10. [CrossRef] [PubMed]

60. Bodenant, M.; Moreau, C.; Sejourné, C.; Auvin, S.; Delval, A.; Cuisset, J.M.; Derambure, P.; Destée, A.; Defebvre, L. Interest of the ketogenic diet in a refractory status epilepticus in adults. *Rev. Neurol.* **2008**, *164*, 194–199. [CrossRef] [PubMed]

61. Wusthoff, C.J.; Kranick, S.M.; Morley, J.F.; Bergqvist, A.G.C. The ketogenic diet in treatment of two adults with prolonged nonconvulsive status epilepticus. *Epilepsia* **2010**, *51*, 1083–1085. [CrossRef] [PubMed]

62. Martikainen, M.H.; Paivarinta, M.; Jaaskelainen, S.; Majamaa, K. Successful treatment of POLG-related mitochondrial epilepsy. *Epileptic Disord.* **2012**, *14*, 438–441. [PubMed]

63. Nam, S.H.; Lee, B.L.; Lee, C.G.; Yu, H.J.; Joo, E.Y.; Lee, J.; Lee, M. The role of ketogenic diet in the treatment of refractory status epilepticus. *Epilepsia* **2011**, *52*, e181–e184. [CrossRef] [PubMed]

64. Strzelczyk, A.; Reif, P.S.; Bauer, S.; Belke, M.; Oertel, W.H.; Knake, S.; Rosenow, F. Intravenous initiation and maintenance of ketogenic diet: Proof of concept in super-refractory status epilepticus. *Seizure* **2013**, *22*, 581–583. [CrossRef] [PubMed]

65. Thakur, K.T.; Probasco, J.C.; Hocker, S.E.; Roehl, K.; Henry, B.; Kossoff, E.H.; Kaplan, P.W.; Geocadin, R.G.; Hartman, A.L.; Venkatesan, A.; et al. Ketogenic diet for adults in super-refractory status epilepticus. *Neurology* **2014**, *82*, 665–670. [CrossRef] [PubMed]

66. Cervenka, M.C.; Hocker, S.E.; Koenig, M.; Bar, B.; Henry-Barron, B.; Kossoff, E.H.; Hartman, A.L.; Probasco, J.C.; Benavides, D.R.; Venkatesan, A.; et al. Phase I/II multicenter ketogenic diet study for adult superrefractory status epilepticus. *Neurology* **2017**, *88*, 938–943. [CrossRef] [PubMed]

67. Shorvon, S.; Ferlisi, M. The outcome of therapies in refractory and super-refractory convulsive status epilepticus and recommendations for therapy. *Brain* **2012**, *135*, 2314–2328. [CrossRef] [PubMed]

68. Wen, P.; Kesari, S. Malignant Gliomas in Adults. *N. Engl. J. Med.* **2008**, *359*, 492–507. [CrossRef] [PubMed]

69. Carlsson, S.K.; Brothers, S.P.; Wahlestedt, C. Emerging treatment strategies for glioblastoma multiforme. *EMBO Mol. Med.* **2014**, *6*, 1359–1370. [CrossRef] [PubMed]

70. Winter, S.F.; Loebel, F.; Dietrich, J. Role of ketogenic metabolic therapy in malignant glioma: A systematic review. *Crit. Rev. Oncol. Hematol.* **2017**, *112*, 41–58. [CrossRef] [PubMed]

71. Warburg, O. On the origin of cancer cells. *Science* **1956**, *123*, 309–314. [CrossRef] [PubMed]

72. Seyfried, T.N.; Flores, R.E.; Poff, A.M.; D'Agostino, D.P. Cancer as a metabolic disease: Implications for novel therapeutics. *Carcinogenesis* **2014**, *35*, 515–527. [CrossRef] [PubMed]

73. Branco, A.F.; Ferreira, A.; Simões, R.F.; Magalhães-Novais, S.; Zehowski, C.; Cope, E.; Silva, A.M.; Pereira, D.; Sardão, V.A.; Cunha-Oliveira, T. Ketogenic diets: From cancer to mitochondrial diseases and beyond. *Eur. J. Clin. Invest.* **2016**, *46*, 285–298. [CrossRef] [PubMed]

74. Jelluma, N. Glucose Withdrawal Induces Oxidative Stress followed by Apoptosis in Glioblastoma Cells but not in Normal Human Astrocytes. *Mol. Cancer Res.* **2006**, *4*, 319–330. [CrossRef] [PubMed]

75. Derr, R.L.; Ye, X.; Islas, M.U.; Desideri, S.; Saudek, C.D.; Grossman, S.A. Association between hyperglycemia and survival in patients with newly diagnosed glioblastoma. *J. Clin. Oncol.* **2009**, *27*, 1082–1086. [CrossRef] [PubMed]

76. McGirt, M.J.; Chaichana, K.L.; Gathinji, M.; Attenello, F.; Than, K.; Ruiz, A.J.; Olivi, A.; Quiñones-Hinojosa, A. Persistent outpatient hyperglycemia is independently associated with decreased survival after primary resection of malignant brain astrocytomas. *Neurosurgery* **2008**, *63*, 286–291. [CrossRef] [PubMed]

77. Mayer, A.; Vaupel, P.; Struss, H.G.; Giese, A.; Stockinger, M.; Schmidberger, H. Ausgeprägt negativer prognostischer Einfluss von hyperglykämischen Episoden während der adjuvanten Radiochemotherapie des Glioblastoma multiforme. *Strahlenther. Onkol.* **2014**, *190*, 933–938. (In German) [CrossRef] [PubMed]

78. Adeberg, S.; Bernhardt, D.; Foerster, R.; Bostel, T.; Koerber, S.A.; Mohr, A.; Koelsche, C.; Rieken, S.; Debus, J. The influence of hyperglycemia during radiotherapy on survival in patients with primary glioblastoma. *Acta Oncol.* **2016**, *55*, 201–207. [CrossRef] [PubMed]

79. Seyfried, T.N.; Sanderson, T.M.; El-Abbadi, M.M.; McGowan, R.; Mukherjee, P. Role of glucose and ketone bodies in the metabolic control of experimental brain cancer. *Br. J. Cancer* **2003**, *89*, 1375–1382. [CrossRef] [PubMed]

80. Marsh, J.; Mukherjee, P.; Seyfried, T.N. Drug/diet synergy for managing malignant astrocytoma in mice: 2-deoxy-D-glucose and the restricted ketogenic diet. *Nutr. Metab.* **2008**, *5*, 1–5. [CrossRef] [PubMed]

81. Abdelwahab, M.G.; Fenton, K.E.; Preul, M.C.; Rho, J.M.; Lynch, A.; Stafford, P.; Scheck, A.C. The ketogenic diet is an effective adjuvant to radiation therapy for the treatment of malignant glioma. *PLoS ONE* **2012**, *7*, 1–7. [CrossRef] [PubMed]

82. Woolf, E.C.; Curley, K.L.; Liu, Q.; Turner, G.H.; Charlton, J.A.; Preul, M.C.; Scheck, A.C. The ketogenic diet alters the hypoxic response and affects expression of proteins associated with angiogenesis, invasive potential and vascular permeability in a mouse glioma model. *PLoS ONE* **2015**, *10*, 1–18. [CrossRef] [PubMed]

83. Woolf, E.C.; Syed, N.; Scheck, A.C. Tumor Metabolism, the Ketogenic Diet and β-Hydroxybutyrate: Novel Approaches to Adjuvant Brain Tumor Therapy. *Front. Mol. Neurosci.* **2016**, *9*, 1–11. [CrossRef] [PubMed]

84. Poff, A.; Koutnik, A.P.; Egan, K.M.; Sahebjam, S.; D'Agostino, D.; Kumar, N.B. Targeting the Warburg effect for cancer treatment: Ketogenic diets for management of glioma. *Semin. Cancer Biol.* **2018**. [CrossRef] [PubMed]

85. Martuscello, R.T.; Vedam-Mai, V.; McCarthy, D.J.; Schmoll, M.E.; Jundi, M.A.; Louviere, C.D.; Griffith, B.G.; Skinner, C.L.; Suslov, O.; Deleyrolle, L.P.; et al. A supplemented high-fat low-carbohydrate diet for the treatment of glioblastoma. *Clin. Cancer Res.* **2016**, *22*, 2482–2495. [CrossRef] [PubMed]

86. Poff, A.M.; Ari, C.; Seyfried, T.N.; D'Agostino, D.P. The Ketogenic Diet and Hyperbaric Oxygen Therapy Prolong Survival in Mice with Systemic Metastatic Cancer. *PLoS ONE* **2013**, *8*. [CrossRef] [PubMed]

87. Cairns, R.A.; Harris, I.S.; Mak, T.W. Regulation of cancer cell metabolism. *Nat. Rev. Cancer* **2011**, *11*, 85–95. [CrossRef] [PubMed]

88. Bowers, L.W.; Rossi, E.L.; O'Flanagan, C.H.; De Graffenried, L.A.; Hursting, S.D. The role of the insulin/IGF system in cancer: Lessons learned from clinical trials and the energy balance-cancer link. *Front. Endocrinol.* **2015**, *6*, 1–16. [CrossRef] [PubMed]

89. Lussier, D.M.; Woolf, E.C.; Johnson, J.L.; Brooks, K.S.; Blattman, J.N.; Scheck, A.C. Enhanced immunity in a mouse model of malignant glioma is mediated by a therapeutic ketogenic diet. *BMC Cancer* **2016**, *16*, 1–10. [CrossRef] [PubMed]

90. Forsythe, C.E.; Phinney, S.D.; Fernandez, M.L.; Quann, E.E.; Wood, R.J.; Bibus, D.M.; Kraemer, W.J.; Feinman, R.D.; Volek, J.S. Comparison of low fat and low carbohydrate diets on circulating fatty acid composition and markers of inflammation. *Lipids* **2008**, *43*, 65–77. [CrossRef] [PubMed]

91. Zuccoli, G.; Marcello, N.; Pisanello, A.; Servadei, F.; Vaccaro, S.; Mukherjee, P.; Seyfried, T.N. Metabolic management of glioblastoma multiforme using standard therapy together with a restricted ketogenic diet: Case Report. *Nutr. Metab.* **2010**, *7*, 1–7. [CrossRef] [PubMed]

92. Champ, C.E.; Palmer, J.D.; Volek, J.S.; Werner-Wasik, M.; Andrews, D.W.; Evans, J.J.; Glass, J.; Kim, L.; Shi, W. Targeting metabolism with a ketogenic diet during the treatment of glioblastoma multiforme. *J. Neurooncol.* **2014**, *117*, 125–131. [CrossRef] [PubMed]

93. Schwartz, K.; Chang, H.T.; Nikolai, M.; Pernicone, J.; Rhee, S.; Olson, K.; Kurniali, P.C.; Hord, N.G.; Noel, M. Treatment of glioma patients with ketogenic diets: Report of two cases treated with an IRB-approved energy-restricted ketogenic diet protocol and review of the literature. *Cancer Metab.* **2015**, *3*, 3. [CrossRef] [PubMed]

94. Rieger, J.; Bähr, O.; Maurer, G.D.; Hattingen, E.; Franz, K.; Brucker, D.; Walenta, S.; Kämmerer, U.; Coy, J.F.; Weller, M.; et al. ERGO: A pilot study of ketogenic diet in recurrent glioblastoma. *Int. J. Oncol.* **2014**, *45*, 1843–1852. [CrossRef] [PubMed]

95. Elsakka, A.M.A.; Bary, M.A.; Abdelzaher, E.; Elnaggar, M.; Kalamian, M.; Mukherjee, P.; Seyfried, T.N. Management of Glioblastoma Multiforme in a Patient Treated With Ketogenic Metabolic Therapy and Modified Standard of Care: A 24-Month Follow-Up. *Front. Nutr.* **2018**, *5*, 1–11. [CrossRef] [PubMed]

96. Martin-McGill, K.J.; Marson, A.G.; Tudur Smith, C.; Jenkinson, M.D. Ketogenic diets as an adjuvant therapy in glioblastoma (the KEATING trial): Study protocol for a randomised pilot study. *Pilot Feasibility Stud.* **2017**, *3*, 1–11. [CrossRef] [PubMed]

97. Swerdlow, R.H. Brain aging, Alzheimer's disease and mitochondria. *Biochim. Biophys. Acta-Mol. Basis Dis.* **2011**, *1812*, 1630–1639. [CrossRef] [PubMed]

98. Wilkins, H.M.; Swerdlow, R.H. Amyloid precursor protein processing and bioenergetics. *Brain Res. Bull.* **2017**, *133*, 71–79. [CrossRef] [PubMed]

99. Castellano, C.A.; Nugent, S.; Paquet, N.; Tremblay, S.; Bocti, C.; Lacombe, G.; Imbeault, H.; Turcotte, E.; Fulop, T.; Cunnane, S. Lower brain 18F-fluorodeoxyglucose uptake but normal 11C-acetoacetate metabolism in mild Alzheimer's disease dementia. *J. Alzheimer's Dis.* **2015**, *43*, 1343–1353. [CrossRef] [PubMed]

100. Winkler, E.A.; Nishida, Y.; Sagare, A.P.; Rege, S.V.; Bell, R.D.; Perlmutter, D.; Sengillo, J.D.; Hillman, S.; Kong, P.; Nelson, A.R.; et al. GLUT1 reductions exacerbate Alzheimer's disease vasculo-neuronal dysfunction and degeneration. *Nat. Neurosci.* **2015**, *18*, 521–530. [CrossRef] [PubMed]

101. Taylor, M.K.; Sullivan, D.K.; Swerdlow, R.H.; Vidoni, E.D.; Morris, J.K.; Mahnken, J.D.; Burns, J.M. A high-glycemic diet is associated with cerebral amyloid burden in cognitively normal older adults. *Am. J. Clin. Nutr.* **2017**, *106*, 1463–1470. [CrossRef] [PubMed]

102. De la Monte, S.M. Insulin Resistance and Neurodegeneration: Progress Towards the Development of New Therapeutics for Alzheimer's Disease. *Drugs* **2017**, *77*, 47–65. [CrossRef] [PubMed]

103. Gaspar, J.M.; Baptista, F.I.; MacEdo, M.P.; Ambrósio, A.F. Inside the Diabetic Brain: Role of Different Players Involved in Cognitive Decline. *ACS Chem. Neurosci.* **2016**, *7*, 131–142. [CrossRef] [PubMed]

104. Nafar, F.; Clarke, J.P.; Mearow, K.M. Coconut oil protects cortical neurons from amyloid beta toxicity by enhancing signaling of cell survival pathways. *Neurochem. Int.* **2017**, *105*, 64–79. [CrossRef] [PubMed]

105. Kashiwaya, Y.; Bergman, C.; Lee, J.H.; Wan, R.; King, M.T.; Mughal, M.R.; Okun, E.; Clarke, K.; Mattson, M.P.; Veech, R.L. A ketone ester diet exhibits anxiolytic and cognition-sparing properties and lessens amyloid and tau pathologies in a mouse model of Alzheimer's disease. *Neurobiol. Aging* **2013**, *34*, 1530–1539. [CrossRef] [PubMed]

106. Studzinski, C.M.; MacKay, W.A.; Beckett, T.L.; Henderson, S.T.; Murphy, M.P.; Sullivan, P.G.; Burnham, W.M.I. Induction of ketosis may improve mitochondrial function and decrease steady-state amyloid-β precursor protein (APP) levels in the aged dog. *Brain Res.* **2008**, *1226*, 209–217. [CrossRef] [PubMed]

107. Kashiwaya, Y.; Takeshima, T.; Mori, N.; Nakashima, K.; Clarke, K.; Veech, R.L. D-b-Hydroxybutyrate protects neurons in models of Alzheimer's and Parkinson's disease. *Proc. Natl. Acad. Sci. USA* **2000**, *97*, 5440–5444. [CrossRef] [PubMed]

108. Van Der Auwera, I.; Wera, S.; Van Leuven, F.; Henderson, S.T. A ketogenic diet reduces amyloid beta 40 and 42 in a mouse model of Alzheimer's disease. *Nutr. Metab.* **2005**, *2*, 1–8. [CrossRef] [PubMed]

109. Yin, J.X.; Maalouf, M.; Han, P.; Zhao, M.; Gao, M.; Dharshaun, T.; Ryan, C.; Whitelegge, J.; Wu, J.; Eisenberg, D.; et al. Ketones block amyloid entry and improve cognition in an Alzheimer's model. *Neurobiol. Aging* **2016**, *39*, 25–37. [CrossRef] [PubMed]

110. Xu, K.; Sun, X.Y.; Eroku, B.O.; Tsipis, C.P.; Puchowicz, M.A.; Lamanna, J.C. Diet-induced ketosis improves cognitive performance in aged rats. In *Oxygen Transport to Tissue XXXI*; Springer: Boston, MA, USA, 2010; Volume 662, pp. 71–75.

111. Reger, M.A.; Henderson, S.T.; Hale, C.; Cholerton, B.; Baker, L.D.; Watson, G.S.; Hyde, K.; Chapman, D.; Craft, S. Effects of β-hydroxybutyrate on cognition in memory-impaired adults. *Neurobiol. Aging* **2004**, *25*, 311–314. [CrossRef]

112. Ota, M.; Matsuo, J.; Ishida, I.; Hattori, K.; Teraishi, T.; Tonouchi, H.; Ashida, K.; Takahashi, T.; Kunugi, H. Effect of a ketogenic meal on cognitive function in elderly adults: Potential for cognitive enhancement. *Psychopharmacology* **2016**, *233*, 3797–3802. [CrossRef] [PubMed]

113. Krikorian, R.; Shidler, M.D.; Dangelo, K.; Couch, S.C.; Benoit, S.C.; Clegg, D.J. Dietary ketosis enhances memory in mild cognitive impairment. *Neurobiol. Aging* **2012**, *33*, 425-e19. [CrossRef] [PubMed]

114. Newport, M.T.; Vanitallie, T.B.; Kashiwaya, Y.; King, M.T.; Veech, R.L. A new way to produce hyperketonemia: Use of ketone ester in a case of Alzheimer's disease. *Alzheimer's Dement.* **2015**, *11*, 99–103. [CrossRef] [PubMed]

115. Taylor, M.K.; Sullivan, D.K.; Mahnken, J.D.; Burns, J.M.; Swerdlow, R.H. Feasibility and efficacy data from a ketogenic diet intervention in Alzheimer's disease. *Alzheimer's Dement. Transl. Res. Clin. Interv.* **2018**, *4*, 28–36. [CrossRef] [PubMed]

116. Henderson, S.T.; Vogel, J.L.; Barr, L.J.; Garvin, F.; Jones, J.J.; Costantini, L.C. Study of the ketogenic agent AC-1202 in mild to moderate Alzheimer's disease: A randomized, double-blind, placebo-controlled, multicenter trial. *Nutr. Metab.* **2009**, *6*, 31. [CrossRef] [PubMed]

117. Rebello, C.J.; Keller, J.N.; Liu, A.G.; Johnson, W.D.; Greenway, F.L. Pilot feasibility and safety study examining the effect of medium chain triglyceride supplementation in subjects with mild cognitive impairment: A randomized controlled trial. *BBA Clin.* **2015**, *3*, 123–125. [CrossRef] [PubMed]

118. Ohnuma, T.; Toda, A.; Kimoto, A.; Takebayashi, Y.; Higashiyama, R.; Tagata, Y.; Ito, M.; Ota, T.; Shibata, N.; Arai, H. Benefits of use and tolerance of, medium-chain triglyceride medical food in the management of Japanese patients with Alzheimer's disease: A prospective, open-label pilot study. *Clin. Interv. Aging* **2016**, *11*, 29–36. [CrossRef] [PubMed]

119. Croteau, E.; Castellano, C.A.; Richard, M.A.; Fortier, M.; Nugent, S.; Lepage, M.; Duchesne, S.; Whittingstall, K.; Turcotte, E.E.; Bocti, C.; et al. Ketogenic Medium Chain Triglycerides Increase Brain Energy Metabolism in Alzheimer's Disease. *J. Alzheimers. Dis.* **2018**, 1–2. [CrossRef] [PubMed]

120. Patel, A.; Pyzik, P.L.; Turner, Z.; Rubenstein, J.E.; Kossoff, E.H. Long-term outcomes of children treated with the ketogenic diet in the past. *Epilepsia* **2010**, *51*, 1277–1282. [CrossRef] [PubMed]

121. Schoeler, N.E.; Cross, J.H. Ketogenic dietary therapies in adults with epilepsy: A practical guide. *Pract. Neurol.* **2016**, *16*, 208–214. [CrossRef] [PubMed]

122. Pinto, A.; Bonucci, A.; Maggi, E.; Corsi, M.; Businaro, R. Anti-Oxidant and Anti-Inflammatory Activity of Ketogenic Diet: New Perspectives for Neuroprotection in Alzheimer's Disease. *Antioxidants* **2018**, *7*, 63. [CrossRef] [PubMed]

123. Tisdale, M. Mechanisms of cancer cachexia. *Physiol. Rev.* **2009**, *89*, 381–410. [CrossRef] [PubMed]

124. Mosek, A.; Natour, H.; Neufeld, M.Y.; Shiff, Y.; Vaisman, N. Ketogenic diet treatment in adults with refractory epilepsy: A prospective pilot study. *Seizure* **2009**, *18*, 30–33. [CrossRef] [PubMed]

125. Cervenka, M.C.; Patton, K.; Eloyan, A.; Henry, B.; Kossoff, E.H. The impact of the modified Atkins diet on lipid profiles in adults with epilepsy. *Nutr. Neurosci.* **2016**, *19*, 131–137. [CrossRef] [PubMed]

126. Mackay, M.T.; Bicknell-Royle, J.; Nation, J.; Humphrey, M.; Harvey, A.S. The ketogenic diet in refractory childhood epilepsy. *J. Paediatr. Child Health* **2005**, *41*, 353–357. [CrossRef] [PubMed]

127. Bergqvist, A.G.C.; Schall, J.I.; Stallings, V.A.; Zemel, B.S. Progressive bone mineral content loss in children with intractable epilepsy treated with the ketogenic diet. *Am. J. Clin. Nutr.* **2008**, *88*, 1678–1684. [CrossRef] [PubMed]

128. Schwartz, K.A.; Noel, M.; Nikolai, M.; Chang, H.T. Investigating the Ketogenic Diet As Treatment for Primary Aggressive Brain Cancer: Challenges and Lessons Learned. *Front. Nutr.* **2018**, *5*. [CrossRef] [PubMed]

129. Mcdonald, T.J.W.; Henry-barron, B.J.; Felton, E.A.; Gutierrez, E.G.; Barnett, J.; Fisher, R.; Lwin, M.; Jan, A.; Vizthum, D.; Kossoff, E.H.; et al. Improving compliance in adults with epilepsy on a modified Atkins diet: A randomized trial. *Seizure Eur. J. Epilepsy* **2018**, *60*, 132–138. [CrossRef] [PubMed]

130. Youngson, N.A.; Morris, M.J.; Ballard, B. The mechanisms mediating the antiepileptic effects of the ketogenic diet and potential opportunities for improvement with metabolism-altering drugs. *Seizure* **2017**, *52*, 15–19. [CrossRef] [PubMed]

![brain sciences logo](brain sciences)

MDPI

Review

Weight Loss Maintenance: Have We Missed the Brain?

Dimitrios Poulimeneas [1], **Mary Yannakoulia** [1], **Costas A. Anastasiou** [1,2]
and Nikolaos Scarmeas [2,3,*]

1   Department of Nutrition and Dietetics, Harokopio University, GR 17676 Athens, Greece;
    dpoul@hua.gr (D.P.); myiannak@hua.gr (M.Y.); acostas@hua.gr (C.A.A.)
2   Eginition Hospital, 1st Neurology Clinic, Department of Social Medicine, Psychiatry and Neurology,
    National and Kapodistrian University of Athens, GR 15772 Athens, Greece
3   Taub Institute for Research in Alzheimer's Disease and the Aging Brain, The Gertrude H. Sergievsky Center,
    Department of Neurology, Columbia University, New York, NY 10027, USA
*   Correspondence: ns257@columbia.edu; Tel.: +30-693-763-8169

Received: 23 July 2018; Accepted: 6 September 2018; Published: 11 September 2018

Abstract: Even though obese individuals often succeed with weight loss, long-term weight loss maintenance remains elusive. Dietary, lifestyle and psychosocial correlates of weight loss maintenance have been researched, yet the nature of maintenance is still poorly understood. Studying the neural processing of weight loss maintainers may provide a much-needed insight towards sustained obesity management. In this narrative review, we evaluate and critically discuss available evidence regarding the food-related neural responses of weight loss maintainers, as opposed to those of obese or lean persons. While research is still ongoing, available data indicate that following weight loss, maintainers exhibit persistent reward related feeling over food, similar to that of obese persons. However, unlike in obese persons, in maintainers, reward-related brain activity appears to be counteracted by subsequently heightened inhibition. These findings suggest that post-dieting, maintainers acquire a certain level of cognitive control which possibly protects them from weight regaining. The prefrontal cortex, as well as the limbic system, encompass key regions of interest for weight loss maintenance, and their contributions to long term successful weight loss should be further explored. Future possibilities and supportive theories are discussed.

Keywords: obesity; weight loss maintenance; maintainers; regainers; neural processing; functional neuroimaging

1. Introduction

Obesity remains a major public health concern at the global level. Excess body weight has been associated with negative effects in multiple organs and body systems [1–3], including the peripheral and the central nervous system. Being obese is associated with several neuropathologies [4], ranging from polyneuropathy [5] and impaired function of various cognitive domains [6] to neurodegenerative diseases, like Alzheimer's disease [7] and other dementias [8,9]. The introduction of neuroimaging in obesity management has further yielded useful information [10]. Obese individuals have been known to exhibit hypothalamic abnormalities [11] hippocampal atrophy [12], and lower brain volume compared to normal-weight or overweight controls [13]. Obesity is associated with both structural and functional alterations in brain areas related to reward anticipation [14,15], inhibition and restraint [16], as well as higher cognitive functioning [17].

On the other hand, weight loss has been found to mitigate neurodegeneration and cognitive decline. In a recent meta-analysis, even modest weight loss (≥2 kg) was associated with improvements in attention, memory, executive functioning and language [18]; larger losses (>10% of initial body weight) have been found to augment cognition in the elderly [19]. The weight loss method may also be of neurological importance. For instance, when compared to behavioral dieters, patients following bariatric surgery have shown enhanced processing in areas related to food motivation (bilateral temporal cortex) [20]. In the same study, dieters showed increased hunger (right medial prefrontal cortex, left precuneus) and self-referent processing, when compared to bariatric patients [20].

However, weight loss is not a milestone, but rather part of a dynamic process. Following weight loss, the weight-reduced individual enters an uneven combat, commonly resulting in weight regain. The available data indicate that the trend of weight regaining in dieters is highly eminent [21] and that ex-obese persons maintain a mere 3–4 kg of their initial weight loss [22], or even less [23]. Hence, researchers have recently focused their interest on individuals who experienced long-term successful weight loss maintenance (SWLM). Several weight control registries have been established worldwide, including the National Weight Control Registry in the US and the MedWeight study in Greece [24–27]. These registries have delineated several factors involved in SWLM, including dietary behaviors [28–31], lifestyle habits [32–36] and psychosocial aspects [37–39], that assist individuals in the longevity of their weight loss.

Despite this knowledge, existing models explain only 30% or less of maintenance variance, leaving the rest unexplained and unexplored [40], suggesting that the nature of maintenance is yet poorly understood. Along these lines, little is known on the sustainability of brain changes during weight loss maintenance. Thus, some research interest has been focused on the neural mechanisms that are involved in weight management and how they could potentiate long term success in post-dieters. In this narrative review, we present and critically discuss weight loss maintenance in relation to the neural processing of individuals with a history of obesity compared to that of lean and/or obese counterparts.

2. Materials and Methods

We searched PubMed for functional neuroimaging studies. Combinations of the following keywords were used: weight loss maintenance/maintainer, weight regain/regainer, functional magnetic resonance imaging (fMRI) or positron emission tomography (PET), neural activity/processing. In addition, the references of the retrieved studies were searched for similar research. Inclusion criteria for this narrative review were (i) publication date from January 2000 till May 2018, (ii) investigations involving human subjects >18 years of age, with absence of psychopathology, (iii) involvement of weight loss maintainers and/or regainers in the sampling. Articles that were involving animal studies, basic neurobiological research, or did not meet the inclusion criteria, were excluded. Our search concluded in 8 studies, and their descriptive information can be found in Table 1. Regional brain activation differences among maintainers, obese and normal weight individuals are summarized in Figure 1.

Brain Sci. **2018**, *8*, 174

Table 1. Descriptive characteristics of the reviewed neuroimaging studies ($n = 8$).

Study, Study Design	Study Population	Weight Loss Maintenance Definition	Imaging Method	Exclusion Criteria	Measures
Del Parigi et al., 2004 [41] Observational Case-Control	11 Maintainers 23 OB 21 NW	Stable weight for ≥3 months, after intentional weight reduction from a BMI ≥ 35 kg/m² to <25 kg/m², through diet and exercise	PET-scan	Not reported	Regional cerebral blood flow at baseline (after a 36-h fast), after tasting and after consuming a satiating liquid meal, in 4 brain regions
Del Parigi et al., 2007 [42] Observational Case-Control	9 Maintainers 20 NW	Stable weight for ≥3 months, after intentional weight reduction from a BMI ≥ 35 kg/m² to <25 kg/m², through diet and exercise	PET-scan	Not reported	Brain response to the sensory experience of food and meal consumption
McCaffery et al., 2009 [43] Observational Case-Control	17 Maintainers 16 OB 18 NW	Maintenance of intentional weight loss ≥13.6 kg, from a maximum BMI ≥ 30 kg/m² to normal BMI, for at least 3 years	fMRI	Medication Left-handedness Neuropathology Psychopathology Standard MRI contradictions	Visual stimuli of low and high calorie foods and non-foods, in a single 8-min run, after a 4-h fast
Hassenstab et al., 2012 [44] Observational Case-Control	17 Maintainers 17 OB 19 NW	Maintenance of intentional weight loss ≥13.6 kg, from a maximum BMI ≥ 30 kg/m² to normal BMI, for at least 3 years	MRI	Medication Neuropathology Psychopathology Standard MRI contradictions	Cortical thickness in 4 *a-priori* set brain regions of the cognitive control network
Sweet et al., 2012 [45] Observational Case-Control	17 Maintainers 14 OB 18 NW	Maintenance of intentional weight loss ≥13.6 kg, from a maximum BMI ≥ 30 kg/m² to a BMI≥18.5 and <27 kg/m², for at least 3 years	fMRI	Medication Left-handedness Neuropathology Psychopathology Standard MRI contradictions	Neurological response during an 1-min orosensory paradigm, after a 4-h fast
Murdaugh et al., 2012 [46] Prospective observation	25 OB, scanned prior and after a 12-week dietary intervention, and on 9-month follow up	Maintenance of weight loss achieved through a 3-month behavioural intervention, 9 months post intervention	fMRI	Left-handedness IQ < 80 Chronic conditions Neuropathology Psychopathology Standard MRI contradictions	Visual stimuli of high-quality color food or non-food photographs
Weygandt et al., 2015 Prospective observation	23 OW and OB, scanned after a 12-week dietary intervention, and on 12-month follow up	Maintenance of weight loss achieved through the dietary intervention	fMRI	Psychopathology Neuropathology	Food related delay-discounting task
Simon et al., 2018 [47] Cross-sectional crossover	17 Maintainers 16 Regainers	Maintenance of weight loss ≥10% of initial body weight, 6 months after a dietary intervention	fMRI	Medication Left-handedness Psychopathology Standard MRI contradictions	Neural processing during two types of incentive delay tasks, during the anticipation and receipt of monetary and/or food-related reward

OW, Overweight; OB, Obese; NW, Normal-Weight; PET, Positron Emission Tomography; MRI, Magnetic Resonance Imaging; fMRI, functional Magnetic Resonance Imaging; BMI: Body Mass Index; IQ: Intelligence quotient.

Neural process	Normal Weight	Maintainers	Obese
Dietary Restraint	☐	☐	
Food Reward	☐	☐	
Inhibitory Control	☐	☐	☐
Hunger processes	☐	☐	
Reward Anticipation	☐	☐	
Reward valuing	☐	☐	☐
Visual processing, Attention		☐	☐
Effective Food Monitoring		☐	☐

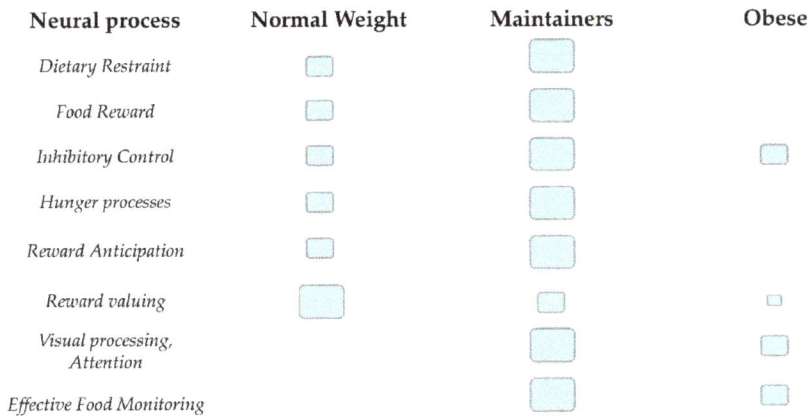

Underlying neural processes for weight loss maintenance proposed by the existing literature. Larger squares indicate higher engagement of suggested neural processes and *vice versa*. When information for all 3 groups is available, comparisons have not necessarily been performed within the framework of the same study (i.e. different study comparing normal weight and mainainers and different study maintainers and obese). Where no square is present indicates that the specific comparison has not been reported in the literature.

Figure 1. A summary of regional brain activity and proposed function in maintainers compared to obese and normal weight individuals.

3. The Neural Background of Weight Loss Maintenance

The notion that cognitive skills may be important for SWLM has interested obesity researchersover the previous decades. In a 2001 study, after a weight reduction intervention, provision of extended care with cognitive component (i.e., problem solving) resulted inmaintenance of >10% of weight loss [48]. What is more, this form of extended care appeared to be more effective for SWLM than relapse prevention training and no extended care. Few years later, Del Parigi and associates [41,42], in two PET-scan studies involving weight loss maintainers, obese and normal weight volunteers recorded that the posterior cingulate and the amygdala were activated after a satiating meal in the obese group, but not in the normal weight or maintainers. In the same study, persistent abnormal responses in the middle insular cortex and the hippocampus of maintainers and obese persons were also reported. These results were of the first to suggest that, being obese or having a history of obesity is associated with greater craving for the coming meal and enhanced memory processing. Additionally, these findings also suggest that when individuals succeed with weight loss, their brain activity differentiates to some extent in relation to their previous obese state.

4. The Interplay of Restraint and Reward Anticipation Brain Regions

As already stated, there seems to exist a strong link between obesity and impaired function of the reward network. The mechanisms could conceivably be explained by the reward-deficiency model [49]. Overweight and obese individuals exhibit greater activation in reward related areas (i.e., insula, amygdala, cingulated gyrus) and reward anticipation areas (anterior cingulate, orbitofrontal cortex) [50]. Higher BMI (Body Mass Index) has been associated with higher activation of reward anticipation and impulsivity regions (anterior cingulated cortex, middle frontal gyrus) in both cross-sectional and prospective studies [51,52]. Finally, a recent systematic review of functional neuroimaging studies suggests consistency in the published research that relates obesity with high reward-related region activation, even after the consumption of a high-calorie meal [50].

In studies involving weight loss maintainers, the picture is similar. However, reward-related processes appear to activate in parallel to a different set of brain areas. Following an orosensory paradigm, Sweet and colleagues [45] observed elevated responses in almost all brain regions examined in maintainers, compared to obese and normal weight controls. As only maintainers exhibited significant reactivity in the left putamen and inferior frontal gyrus (areas associated with food reward and inhibitory control, respectively), the authors hypothesized that maintainers exhibited greater reward expectations during the orosensory stimulation, but responded with greater restraint. Greater restraint in maintainers has been previously reported also in an observational study using visual stimuli of low and high calorie food pictures [43]. Compared to obese and normal weight controls, the maintainers seemed to experience greater inhibition and restraint, as they showed greater activation in the left superior frontal region of the brain, this "inhibitory" activation of the prefrontal cortex that maintainers experience is similar to the pattern found in normal-weight individuals [53,54].

Even though the activation of the inferior frontal gyrus during various stimuli has been associated with SWLM, the prospective study by Murdaugh et al. did not support it [46]: weight maintenance 9 months following a dietary intervention was associated with decreased post-treatment activity in the insula and the putamen, as well as the midbrain/thalamus and the inferior frontal gyrus. These contradictory findings may be partly explained by the different methodologies and selection criteria used in various studies.

According to Sweet et al. [45], when maintainers are provided with a food stimulus, their brain reaction follows a pattern of elevated reward expectation, yet consequently greater inhibitory control. In a prospective study, greater impulse control (as expressed by the activation of the dorsolateral prefrontal cortex), immediately after a 12-week behavioral weight loss intervention, was found to be predictive of SWLM in the 1-year follow up [55]. Additionally, a recent study exploring cortical thickness as a surrogate marker of cognitive control, concluded that maintainers tended to have greater cortical thickness than obese controls (although the trend did not pass the significance threshold) [44]. Taken together these findings may imply that, in the post-dieting period, maintainers could exhibit cognitive changes, which counteract their food reward-related neural circuits and possibly protect them against weight regain.

5. Executive Functions Driving Maintenance

As the prefrontal cortex may play a pivotal role in protecting against weight regaining the potential involvement of executive functioning has been investigated. Executive functions, a neuropsychological trait regulated by the prefrontal cortex [56], mediate processes of "how" choices are made and established [57] and may moderate the relationship between eating intention and behavior [58]. Higher executive functions have been linked to weight loss maintenance during the 2-year post-procedural period of bariatric surgery patients [59]. Similarly, in a functional neuroimaging study of people post obesity surgery, greater utilization of the executive control circuitry of resistance to palatable food cues was associated with more successful maintenance [51]. Thus, manipulation of executive functions through cognitive training, towards enhancing and optimizing function of the prefrontal cortex, may hold promising aspects for long term weight loss maintenance [60,61].

6. Clinical Implications and Future Possibilities

In summary, while evidence is still insufficient, following weight loss, maintainers acquire a certain level of cognition, reacting with heightened inhibition against food cues, to compensate for elevated reward related feeling over food. The differences in the neural activity of weight loss between maintainers and obese individuals indicate a persistent imbalance between hedonic and homeostatic food ingestion following weight loss. The prefrontal cortex, as well as the limbic system are key regions of interest for weight loss maintenance and their contributions to long term successful weight loss should be further explored.

Although some evidence regarding the neural responses of weight loss maintainers has been accumulated, there are major caveats in the comparison and reproducibility of the results from the neuroimaging studies [50]. As indicated in Table 1, the measures of acquisition have not been systematic, there is no standard approach in methodology (scan acquisition, inclusion criteria etc.) while the studies so far have commonly used small-sized, convenient samples. What is more, not all regions that activate differently in maintainers in comparison to obese or normal weight individuals have been assigned a proposed function, thus inadequately profiling neural involvement in weight loss maintenance (Figure 1). Future research should address these issues and exploit standardized approaches in larger population groups.

Following weight loss, maintainers exhibit a transitioning period during which they show brain similarities to both obese and normal weight persons. The duration of the transitioning period, if finite, is yet unknown: it begins after weight loss and may span beyond 3 years of maintenance. New studies should examine the brain regions of people who have achieved to maintain their weight loss for prolonged periods of time (i.e., $\geq$5 years of maintenance) that activate, when exposed to food cues. In addition, neuroimaging research should directly compare weight loss maintainers with individuals that regained their weight loss shortly after the dieting period, or maintained it for more than a year, but regained the loss thereafter (ex-maintainers).

The interplay of restraint and reward for weight loss maintenance could be explored from different perspectives. Multistep cognitive behavioral treatment in obese patients has produced promising results in enhancing dietary restraint [62]. Our understanding of neural restraint in weight loss maintenance would be enhanced by studies examining not only the neural activity of selected brain regions, but also their functional connectivity in the resting state. For example, resting state activity of the middle temporal gyrus has been shown to correlate with dietary restraint [63]; this association was supposed to reflect the middle temporal gurus' connectivity with frontal regions involved in inhibitory processes [64]. Exploring the sensory experience of food, as well as the impact of food architecture may also be of importance in addressing food reward [65]. Personality traits, such as persistence [66], may favor weight loss maintenance, and their implications in the abovementioned interplay should be further researched. Last, examining the role of neuropeptides with known homeostatic properties, prominent in the limbic system and the prefrontal cortex, such as orexin [67,68], may withhold therapeutic targets for long-term obesity management [69].

To succeed with weight loss maintenance, post obese individuals are required to exercise morethan a dieter [70], and to adhere to a low-calorie diet [28], 300–400 kcal lower of that expected of their body mass [71], to compensate for decreased energy expenditure and persistent physiological adaptations that favor weight regain [72].Considering the brain similarities of the maintainers to both the obese and normal-weight persons, we hypothesize that people with previous history of obesity, even if presented with normal BMI, should not be treated as normal-weight, but rather as ex-obese or weight-reduced individuals. This postulation is strengthened by research that focuses beyond behavior or neuroimaging. For instance, epigenetic DNA methylation patterns of maintainers has been found to more closely resemble that of normal weight than obese controls [73]. If supported by future research, this hypothesis may provide a paradigm shift for clinicians and obesity researchers, so as to more thoroughly profile and prevent weight regaining.

Author Contributions: Writing original draft: D.P. Writing review and editing: M.Y., C.A.A. and N.S. Conceptualization: N.S., M.Y.

Funding: This research received no external funding. Costas A. Anastasiou has received financial support from the Greek State Scholarships Foundation (MIS: 5001552).

Conflicts of Interest: Mary Yannakoulia was the PI of the MedWeight study (2012–2015), that was partially funded by the Coca Cola Foundation (KA 221). Nikolaos Scarmeas reportspersonal fees from Merck Consumer Health. Dimitrios Poulimeneas and Costas A. Anastasiou declare no conflicts of interest.

References

1. Ndumele, C.E.; Matsushita, K.; Lazo, M.; Bello, N.; Blumenthal, R.S.; Gerstenblith, G.; Nambi, V.; Ballantyne, C.M.; Solomon, S.D.; Selvin, E.; et al. Obesity and subtypes of incident cardiovascular disease. *J. Am. Heart. Assoc.* **2016**, *5*, e003921. [CrossRef] [PubMed]
2. Langenberg, C.; Sharp, S.J.; Schulze, M.B.; Rolandsson, O.; Overvad, K.; Forouhi, N.G.; Spranger, J.; Drogan, D.; Huerta, J.M.; Arriola, L.; et al. Long-term risk of incident type 2 diabetes and measures of overall and regional obesity: The epic-interact case-cohort study. *PLoS Med.* **2012**, *9*, e1001230.

3. Kyrgiou, M.; Kalliala, I.; Markozannes, G.; Gunter, M.J.; Paraskevaidis, E.; Gabra, H.; Martin-Hirsch, P.; Tsilidis, K.K. Adiposity and cancer at major anatomical sites: Umbrella review of the literature. *BMJ* **2017**, *356*, j477. [CrossRef] [PubMed]

4. Monda, V.; La Marra, M.; Perrella, R.; Caviglia, G.; Iavarone, A.; Chieffi, S.; Messina, G.; Carotenuto, M.; Monda, M.; Messina, A. Obesity and brain illness: From cognitive and psychological evidences to obesity paradox. *Diabetes Metab. Syndr. Obes.* **2017**, *10*, 473–479. [CrossRef] [PubMed]

5. Sone, H.; Mizuno, S.; Yamada, N. Vascular risk factors and diabetic neuropathy. *N. Engl. J. Med.* **2005**, *352*, 1925–1927. [PubMed]

6. Prickett, C.; Brennan, L.; Stolwyk, R. Examining the relationship between obesity and cognitive function: A systematic literature review. *Obes. Res. Clin. Pract.* **2015**, *9*, 93–113. [CrossRef] [PubMed]

7. Hazar, N.; Seddigh, L.; Rampisheh, Z.; Nojomi, M. Population attributable fraction of modifiable risk factors for Alzheimer disease: A systematic review of systematic reviews. *Iran. J. Neurol.* **2016**, *15*, 164–172. [PubMed]

8. Albanese, E.; Launer, L.J.; Egger, M.; Prince, M.J.; Giannakopoulos, P.; Wolters, F.J.; Egan, K. Body mass index in midlife and dementia: Systematic review and meta-regression analysis of 589,649 men and women followed in longitudinal studies. *Alzheimers Dement. (Amst.)* **2017**, *8*, 165–178. [CrossRef] [PubMed]

9. Pedditzi, E.; Peters, R.; Beckett, N. Corrigenda: Corrigendum to the risk of overweight/obesity in mid-life and late life for the development of dementia: A systematic review and meta-analysis of longitudinal studies'. *Age Ageing* **2016**, *45*, 740. [CrossRef] [PubMed]

10. Behary, P.; Miras, A.D. Brain responses to food and weight loss. *Exp. Physiol.* **2014**, *99*, 1121–1127. [CrossRef] [PubMed]

11. Williams, L.M. Hypothalamic dysfunction in obesity. *Proc. Nutr. Soc.* **2012**, *71*, 521–533. [CrossRef] [PubMed]

12. Cherbuin, N.; Sargent-Cox, K.; Fraser, M.; Sachdev, P.; Anstey, K.J. Being overweight is associated with hippocampal atrophy: The path through life study. *Int. J. Obes. (Lond.)* **2015**, *39*, 1509–1514. [CrossRef] [PubMed]

13. Gunstad, J.; Paul, R.H.; Cohen, R.A.; Tate, D.F.; Spitznagel, M.B.; Grieve, S.; Gordon, E. Relationship between body mass index and brain volume in healthy adults. *Int. J. Neurosci.* **2008**, *118*, 1582–1593. [CrossRef] [PubMed]

14. Gupta, A.; Mayer, E.A.; Labus, J.S.; Bhatt, R.R.; Ju, T.; Love, A.; Bal, A.; Tillisch, K.; Naliboff, B.; Sanmiguel, C.P.; et al. Sex commonalities and differences in obesity-related alterations in intrinsic brain activity and connectivity. *Obesity (Silver Spring)* **2018**, *26*, 340–350. [CrossRef] [PubMed]

15. Morris, M.J.; Beilharz, J.E.; Maniam, J.; Reichelt, A.C.; Westbrook, R.F. Why is obesity such a problem in the 21st century? The intersection of palatable food, cues and reward pathways, stress, and cognition. *Neurosci. Biobehav. Rev.* **2015**, *58*, 36–45. [CrossRef] [PubMed]

16. Jokela, M.; Hintsanen, M.; Hakulinen, C.; Batty, G.D.; Nabi, H.; Singh-Manoux, A.; Kivimaki, M. Association of personality with the development and persistence of obesity: A meta-analysis based on individual-participant data. *Obes. Rev.* **2013**, *14*, 315–323. [CrossRef] [PubMed]

17. Horstmann, A. It wasn't me; it was my brain-obesity-associated characteristics of brain circuits governing decision-making. *Physiol. Behav.* **2017**, *176*, 125–133. [CrossRef] [PubMed]

18. Veronese, N.; Facchini, S.; Stubbs, B.; Luchini, C.; Solmi, M.; Manzato, E.; Sergi, G.; Maggi, S.; Cosco, T.; Fontana, L. Weight loss is associated with improvements in cognitive function among overweight and obese people: A systematic review and meta-analysis. *Neurosci. Biobehav. Rev.* **2017**, *72*, 87–94. [CrossRef] [PubMed]

19. Napoli, N.; Shah, K.; Waters, D.L.; Sinacore, D.R.; Qualls, C.; Villareal, D.T. Effect of weight loss, exercise, or both on cognition and quality of life in obese older adults. *Am. J. Clin. Nutr.* **2014**, *100*, 189–198. [CrossRef] [PubMed]

20. Bruce, A.S.; Bruce, J.M.; Ness, A.R.; Lepping, R.J.; Malley, S.; Hancock, L.; Powell, J.; Patrician, T.M.; Breslin, F.J.; Martin, L.E.; et al. A comparison of functional brain changes associated with surgical versus behavioral weight loss. *Obesity (Silver Spring)* **2014**, *22*, 337–343. [CrossRef] [PubMed]

21. Montesi, L.; El Ghoch, M.; Brodosi, L.; Calugi, S.; Marchesini, G.; Dalle Grave, R. Long-term weight loss maintenance for obesity: A multidisciplinary approach. *Diabetes Metab. Syndr. Obes.* **2016**, *9*, 37–46. [PubMed]

22. Franz, M.J.; VanWormer, J.J.; Crain, A.L.; Boucher, J.L.; Histon, T.; Caplan, W.; Bowman, J.D.; Pronk, N.P. Weight-loss outcomes: A systematic review and meta-analysis of weight-loss clinical trials with a minimum 1-year follow-up. *J. Am. Diet. Assoc.* **2007**, *107*, 1755–1767. [CrossRef] [PubMed]

23. Dombrowski, S.U.; Knittle, K.; Avenell, A.; Araujo-Soares, V.; Sniehotta, F.F. Long term maintenance of weight loss with non-surgical interventions in obese adults: Systematic review and meta-analyses of randomised controlled trials. *BMJ* **2014**, *348*, g2646. [CrossRef] [PubMed]

24. Klem, M.L.; Wing, R.R.; McGuire, M.T.; Seagle, H.M.; Hill, J.O. A descriptive study of individuals successful at long-term maintenance of substantial weight loss. *Am. J. Clin. Nutr.* **1997**, *66*, 239–246. [CrossRef] [PubMed]

25. Santos, I.; Vieira, P.N.; Silva, M.N.; Sardinha, L.B.; Teixeira, P.J. Weight control behaviors of highly successful weight loss maintainers: The portuguese weight control registry. *J. Behav. Med.* **2017**, *40*, 366–371. [CrossRef] [PubMed]

26. Soini, S.; Mustajoki, P.; Eriksson, J.G. Weight loss methods and changes in eating habits among successful weight losers. *Ann. Med.* **2016**, *48*, 76–82. [CrossRef] [PubMed]

27. Karfopoulou, E.; Anastasiou, C.A.; Hill, J.O.; Yannakoulia, M. The medweight study: Design and preliminary results. *Mediterr. J. Nutr. Metab.* **2014**, *7*, 201–210.

28. Shick, S.M.; Wing, R.R.; Klem, M.L.; McGuire, M.T.; Hill, J.O.; Seagle, H. Persons successful at long-term weight loss and maintenance continue to consume a low-energy, low-fat diet. *J. Am. Diet. Assoc.* **1998**, *98*, 408–413. [CrossRef]

29. Brikou, D.; Zannidi, D.; Karfopoulou, E.; Anastasiou, C.A.; Yannakoulia, M. Breakfast consumption and weight-loss maintenance: Results from the medweight study. *Br. J. Nutr.* **2016**, *115*, 2246–2251. [CrossRef] [PubMed]

30. Wyatt, H.R.; Grunwald, G.K.; Mosca, C.L.; Klem, M.L.; Wing, R.R.; Hill, J.O. Long-term weight loss and breakfast in subjects in the national weight control registry. *Obes. Res.* **2002**, *10*, 78–82. [CrossRef] [PubMed]

31. Karfopoulou, E.; Brikou, D.; Mamalaki, E.; Bersimis, F.; Anastasiou, C.A.; Hill, J.O.; Yannakoulia, M. Dietary patterns in weight loss maintenance: Results from the medweight study. *Eur. J. Nutr.* **2017**, *56*, 991–1002. [CrossRef] [PubMed]

32. Catenacci, V.A.; Ogden, L.G.; Stuht, J.; Phelan, S.; Wing, R.R.; Hill, J.O.; Wyatt, H.R. Physical activity patterns in the national weight control registry. *Obesity (Silver Spring)* **2008**, *16*, 153–161. [CrossRef] [PubMed]

33. Catenacci, V.A.; Odgen, L.; Phelan, S.; Thomas, J.G.; Hill, J.; Wing, R.R.; Wyatt, H. Dietary habits and weight maintenance success in high versus low exercisers in the national weight control registry. *J. Phys. Act. Health* **2014**, *11*, 1540–1548. [CrossRef] [PubMed]

34. Phelan, S.; Wyatt, H.R.; Hill, J.O.; Wing, R.R. Are the eating and exercise habits of successful weight losers changing? *Obesity (Silver Spring)* **2006**, *14*, 710–716. [CrossRef] [PubMed]

35. Yannakoulia, M.; Anastasiou, C.A.; Karfopoulou, E.; Pehlivanidis, A.; Panagiotakos, D.B.; Vgontzas, A. Sleep quality is associated with weight loss maintenance status: The medweight study. *Sleep Med.* **2017**, *34*, 242–245. [CrossRef] [PubMed]

36. Ross, K.M.; Graham Thomas, J.; Wing, R.R. Successful weight loss maintenance associated with morning chronotype and better sleep quality. *J. Behav. Med.* **2016**, *39*, 465–471. [CrossRef] [PubMed]

37. Klem, M.L.; Wing, R.R.; McGuire, M.T.; Seagle, H.M.; Hill, J.O. Psychological symptoms in individuals successful at long-term maintenance of weight loss. *Health Psychol.* **1998**, *17*, 336–345. [CrossRef] [PubMed]

38. Karfopoulou, E.; Anastasiou, C.A.; Avgeraki, E.; Kosmidis, M.H.; Yannakoulia, M. The role of social support in weight loss maintenance: Results from the medweight study. *J. Behav. Med.* **2016**, *39*, 511–518. [CrossRef] [PubMed]

39. Anastasiou, C.A.; Fappa, E.; Karfopoulou, E.; Gkza, A.; Yannakoulia, M. Weight loss maintenance in relation to locus of control: The medweight study. *Behav. Res. Ther.* **2015**, *71*, 40–44. [CrossRef] [PubMed]

40. Teixeira, P.J.; Silva, M.N.; Coutinho, S.R.; Palmeira, A.L.; Mata, J.; Vieira, P.N.; Carraca, E.V.; Santos, T.C.; Sardinha, L.B. Mediators of weight loss and weight loss maintenance in middle-aged women. *Obesity (Silver Spring)* **2010**, *18*, 725–735. [CrossRef] [PubMed]

41. DelParigi, A.; Chen, K.; Salbe, A.D.; Hill, J.O.; Wing, R.R.; Reiman, E.M.; Tataranni, P.A. Persistence of abnormal neural responses to a meal in postobese individuals. *Int. J. Obes. Relat. Metab. Disord.* **2004**, *28*, 370–377. [CrossRef] [PubMed]

42. DelParigi, A.; Chen, K.; Salbe, A.D.; Hill, J.O.; Wing, R.R.; Reiman, E.M.; Tataranni, P.A. Successful dieters have increased neural activity in cortical areas involved in the control of behavior. *Int. J. Obes. (Lond.)* **2007**, *31*, 440–448. [CrossRef] [PubMed]

43. McCaffery, J.M.; Haley, A.P.; Sweet, L.H.; Phelan, S.; Raynor, H.A.; Del Parigi, A.; Cohen, R.; Wing, R.R. Differential functional magnetic resonance imaging response to food pictures in successful weight–loss maintainers relative to normal–weight and obese controls. *Am. J. Clin. Nutr.* **2009**, *90*, 928–934. [CrossRef] [PubMed]

44. Hassenstab, J.J.; Sweet, L.H.; Del Parigi, A.; McCaffery, J.M.; Haley, A.P.; Demos, K.E.; Cohen, R.A.; Wing, R.R. Cortical thickness of the cognitive control network in obesity and successful weight loss maintenance: A preliminary mri study. *Psychiatry Res.* **2012**, *202*, 77–79. [CrossRef] [PubMed]

45. Sweet, L.H.; Hassenstab, J.J.; McCaffery, J.M.; Raynor, H.A.; Bond, D.S.; Demos, K.E.; Haley, A.P.; Cohen, R.A.; Del Parigi, A.; Wing, R.R. Brain response to food stimulation in obese, normal weight, and successful weight loss maintainers. *Obesity (Silver Spring)* **2012**, *20*, 2220–2225. [CrossRef] [PubMed]

46. Murdaugh, D.L.; Cox, J.E.; Cook, E.W., 3rd; Weller, R.E. Fmri reactivity to high–calorie food pictures predicts short-and long-term outcome in a weight-loss program. *Neuroimage* **2012**, *59*, 2709–2721. [CrossRef] [PubMed]

47. Simon, J.J.; Becker, A.; Sinno, M.H.; Skunde, M.; Bendszus, M.; Preissl, H.; Enck, P.; Herzog, W.; Friederich, H.C. Neural food reward processing in successful and unsuccessful weight maintenance. *Obesity (Silver Spring)* **2018**, *26*, 895–902. [CrossRef] [PubMed]

48. Perri, M.G.; Nezu, A.M.; McKelvey, W.F.; Shermer, R.L.; Renjilian, D.A.; Viegener, B.J. Relapse prevention training and problem–solving therapy in the long–term management of obesity. *J. Consult. Clin. Psychol.* **2001**, *69*, 722–726. [CrossRef] [PubMed]

49. Volkow, N.D.; Wang, G.J.; Baler, R.D. Reward, dopamine and the control of food intake: Implications for obesity. *Trends Cogn. Sci.* **2011**, *15*, 37–46. [CrossRef] [PubMed]

50. Pursey, K.M.; Stanwell, P.; Callister, R.J.; Brain, K.; Collins, C.E.; Burrows, T.L. Neural responses to visual food cues according to weight status: A systematic review of functional magnetic resonance imaging studies. *Front. Nutr.* **2014**, *1*, 7. [CrossRef] [PubMed]

51. Goldman, R.L.; Canterberry, M.; Borckardt, J.J.; Madan, A.; Byrne, T.K.; George, M.S.; O'Neil, P.M.; Hanlon, C.A. Executive control circuitry differentiates degree of success in weight loss following gastric-bypass surgery. *Obesity (Silver Spring)* **2013**, *21*, 2189–2196. [CrossRef] [PubMed]

52. Bruce, J.M.; Hancock, L.; Bruce, A.; Lepping, R.J.; Martin, L.; Lundgren, J.D.; Malley, S.; Holsen, L.M.; Savage, C.R. Changes in brain activation to food pictures after adjustable gastric banding. *Surg. Obes. Relat. Dis.* **2012**, *8*, 602–608. [CrossRef] [PubMed]

53. Le, D.S.; Pannacciulli, N.; Chen, K.; Salbe, A.D.; Del Parigi, A.; Hill, J.O.; Wing, R.R.; Reiman, E.M.; Krakoff, J. Less activation in the left dorsolateral prefrontal cortex in the reanalysis of the response to a meal in obese than in lean women and its association with successful weight loss. *Am. J. Clin. Nutr.* **2007**, *86*, 573–579. [CrossRef] [PubMed]

54. Le, D.S.; Pannacciulli, N.; Chen, K.; Del Parigi, A.; Salbe, A.D.; Reiman, E.M.; Krakoff, J. Less activation of the left dorsolateral prefrontal cortex in response to a meal: A feature of obesity. *Am. J. Clin. Nutr.* **2006**, *84*, 725–731. [CrossRef] [PubMed]

55. Weygandt, M.; Mai, K.; Dommes, E.; Ritter, K.; Leupelt, V.; Spranger, J.; Haynes, J.D. Impulse control in the dorsolateral prefrontal cortex counteracts post-diet weight regain in obesity. *Neuroimage* **2015**, *109*, 318–327. [CrossRef] [PubMed]

56. Val-Laillet, D.; Aarts, E.; Weber, B.; Ferrari, M.; Quaresima, V.; Stoeckel, L.E.; Alonso-Alonso, M.; Audette, M.; Malbert, C.H.; Stice, E. Neuroimaging and neuromodulation approaches to study eating behavior and prevent and treat eating disorders and obesity. *Neuroimage Clin.* **2015**, *8*, 1–31. [CrossRef] [PubMed]

57. Lezak, M.D.; Howieson, D.B.; Loring, D.W.; Hannay, H.J.; Fischer, J.S. *Neuropsychological Assessment*, 4th ed.; Oxford University Press: New York, NY, USA, 2012.

58. Nederkoorn, C.; Houben, K.; Hofmann, W.; Roefs, A.; Jansen, A. Control yourself or just eat what you like? Weight gain over a year is predicted by an interactive effect of response inhibition and implicit preference for snack foods. *Health Psychol.* **2010**, *29*, 389–393. [CrossRef] [PubMed]

59. Spitznagel, M.B.; Alosco, M.; Strain, G.; Devlin, M.; Cohen, R.; Paul, R.; Crosby, R.D.; Mitchell, J.E.; Gunstad, J. Cognitive function predicts 24-month weight loss success following bariatric surgery. *Surg. Obes. Relat. Dis.* **2013**, *9*, 765–770. [CrossRef] [PubMed]

60. Gettens, K.M.; Gorin, A.A. Executive function in weight loss and weight loss maintenance: A conceptual review and novel neuropsychological model of weight control. *J. Behav. Med.* **2017**, *40*, 687–701. [CrossRef] [PubMed]

61. Jones, A.; Hardman, C.A.; Lawrence, N.; Field, M. Cognitive training as a potential treatment for overweight and obesity: A critical review of the evidence. *Appetite* **2018**, *124*, 50–67. [CrossRef] [PubMed]

62. Dalle Grave, R.; Sartirana, M.; El Ghoch, M.; Calugi, S. Personalized multistep cognitive behavioral therapy for obesity. *Diabetes Metab. Syndr. Obes.* **2017**, *10*, 195–206. [CrossRef] [PubMed]

63. Zhao, J.; Li, M.; Zhang, Y.; Song, H.; Von Deneen, K.M.; Shi, Y.; Liu, Y.; He, D. Intrinsic brain subsystem associated with dietary restraint, disinhibition and hunger: An fmri study. *Brain Imaging Behav.* **2017**, *11*, 264–277. [CrossRef] [PubMed]

64. Olivo, G.; Zhou, W.; Sundbom, M.; Zhukovsky, C.; Hogenkamp, P.; Nikontovic, L.; Stark, J.; Wiemerslage, L.; Larsson, E.-M.; Benedict, C.; et al. Resting-state brain connectivity changes in obese women after roux-en-y gastric bypass surgery: A longitudinal study. *Sci. Rep.* **2017**, *7*, 6616. [CrossRef] [PubMed]

65. Pandit, R.; Mercer, J.G.; Overduin, J.; La Fleur, S.E.; Adan, R.A. Dietary factors affect food reward and motivation to eat. *Obes. Facts* **2012**, *5*, 221–242. [CrossRef] [PubMed]

66. Dalle Grave, R.; Calugi, S.; El Ghoch, M. Are personality characteristics as measured by the temperament and character inventory (TCI) associated with obesity treatment outcomes? A systematic review. *Curr. Obes. Rep.* **2018**, *7*, 27–36. [CrossRef] [PubMed]

67. Monda, V.; Salerno, M.; Sessa, F.; Bernardini, R.; Valenzano, A.; Marsala, G.; Zammit, C.; Avola, R.; Carotenuto, M.; Messina, G.; et al. Functional changes of orexinergic reaction to psychoactive substances. *Mol. Neurobiol.* **2018**, *55*, 6362–6368. [CrossRef] [PubMed]

68. Chieffi, S.; Carotenuto, M.; Monda, V.; Valenzano, A.; Villano, I.; Precenzano, F.; Tafuri, D.; Salerno, M.; Filippi, N.; Nuccio, F.; et al. Orexin system: The key for a healthy life. *Front. Physiol.* **2017**, *8*, 357. [CrossRef] [PubMed]

69. Boughton, C.K.; Murphy, K.G. Can neuropeptides treat obesity? A review of neuropeptides and their potential role in the treatment of obesity. *Br. J. Pharmacol.* **2013**, *170*, 1333–1348. [CrossRef] [PubMed]

70. Jensen, M.D.; Ryan, D.H.; Apovian, C.M.; Ard, J.D.; Comuzzie, A.G.; Donato, K.A.; Hu, F.B.; Hubbard, V.S.; Jakicic, J.M.; Kushner, R.F.; et al. 2013 AHA/ACC/TOS guideline for the management of overweight and obesity in adults: A report of the American college of cardiology/American heart association task force on practice guidelines and the obesity society. *Circulation* **2014**, *129*, S102–S138. [CrossRef] [PubMed]

71. Hinkle, W.; Cordell, M.; Leibel, R.; Rosenbaum, M.; Hirsch, J. Effects of reduced weight maintenance and leptin repletion on functional connectivity of the hypothalamus in obese humans. *PLoS ONE* **2013**, *8*, e59114. [CrossRef] [PubMed]

72. Greenway, F.L. Physiological adaptations to weight loss and factors favouring weight regain. *Int. J. Obes. (Lond.)* **2015**, *39*, 1188–1196. [CrossRef] [PubMed]

73. Huang, Y.-T.; Maccani, J.Z.J.; Hawley, N.L.; Wing, R.R.; Kelsey, K.T.; McCaffery, J.M. Epigenetic patterns in successful weight loss maintainers: A pilot study. *Int. J. Obes. (Lond.)* **2015**, *39*, 865–868. [CrossRef] [PubMed]

MDPI

St. Alban-Anlage 66

4052 Basel

Switzerland

Tel. +41 61 683 77 34

Fax +41 61 302 89 18

www.mdpi.com

Brain Sciences Editorial Office

E-mail: brainsci@mdpi.com

www.mdpi.com/journal/brainsci

www.ingramcontent.com/pod-product-compliance
Lightning Source LLC
Chambersburg PA
CBHW051917210326
41597CB00033B/6174